January 20, 1984

To Peter,
with best wishes
and many thanks
for all your help.
Harold and Joann

Risk Analysis and Decision Processes

Risk Analysis and Decision Processes

The Siting of Liquefied Energy Gas Facilities in Four Countries

Howard C. Kunreuther and Joanne Linnerooth

John Lathrop
Hermann Atz
Sally Macgill
Christoph Mandl
Michiel Schwarz
Michael Thompson

With a Foreword by Mary Douglas

With 29 Figures and 35 Tables

Springer-Verlag
Berlin Heidelberg New York Tokyo 1983

International Institute for Applied Systems Analysis (IIASA)
A-2361 Laxenburg, Austria

Howard C. Kunreuther
Decision Sciences Department, The Wharton School, University of Pennsylvania, Philadelphia, PA 19104, USA

Joanne Linnerooth
International Institute for Applied Systems Analysis,
A-2361 Laxenburg, Austria

John Lathrop
Woodward Clyde Consultants, 3 Embarcadero Center, San Francisco, CA 94111, USA

Hermann Atz
International Institute for Applied Systems Analysis,
A-2361 Laxenburg, Austria

Sally Macgill
School of Geography, The University of Leeds, Leeds LS2 9JT, UK

Christoph Mandl
Elin-Union AG für Elektrische Industrie, Penzingerstraße 76,
A-1141 Wien, Austria

Michiel Schwarz
Department of Social and Economic Studies, Imperial College of Science and Technology, University of London, London SW7 2AZ, UK

Michael Thompson
International Institute for Applied Systems Analysis,
A-2361 Laxenburg, Austria

ISBN 3-540-12804-2 Springer-Verlag Berlin Heidelberg New York Tokyo
ISBN 0-387-12804-2 Springer-Verlag New York Heidelberg Berlin Tokyo

Library of Congress Cataloging in Publication Data
Kunreuther, Howard. Risk analysis and decision processes.
"International Institute for Applied Systems Analysis" – T. p. verso. Bilbliography: p. Includes index.
1. Liquefied natural gas industry – Location – Decision making – Case studies.
2. Liquefied petroleum gas industry – Location – Decision making – Case studies.
3. Liquefied natural gas industry – Environmental aspects.
4. Liquefied petroleum gas industry – Environmental aspects.
5. Risk.
I. Linnerooth, Joanne, 1947–.
II. International Institute for Applied Systems Analysis.
III. Title. HD9581.2.L572K86 1983 338.6'042 83-20044

Typeset in Great Britain by The Alden Press, London. Printing and Bookbinding: K. Triltsch, Würzburg
2060/3020 - 543210

To IIASA
and
To our families
Sylvia, Laura, and Joel,
Kurt and Jan

Foreword

For old and new studies in decision making and risk analysis, this book should stand at the watershed. Studies of conflict resolution and public policy will surely now have to take account of the model investigation provided by the IIASA team, and many things will not be the same again. This is a report of inquiries into the siting of liquefied energy gas (LEG) facilities in the Federal Republic of Germany, the Netherlands, the United Kingdom, and the United States. The risks of transporting this highly combustible stuff, and the economic benefits of being able to bring a natural energy source from one side of the globe to the other, holding it, and piping it out as needed, make LEG a model case for studying the public response to dangerous technology. The dangers of LEG are different from those of nuclear power, for instance, where the response too often becomes entangled with the fear of nuclear war. The dangers of LEG include uncontrollable explosions, rather than insidious contamination. But the degree of dangerousness is very much of the same order as that of nuclear power, and is at least as difficult to assess.

In four different countries the constitutional procedures involved in obtaining approval of nuclear or LEG facilities are on record. The four case histories here are a model for comparative study of conflict resolution. The period over which the negotiations developed is much the same. The political structures, economic pressures, and local representative institutions differ, however, and in each case the events unfolded in slightly different ways. To bring together in one strong framework case histories, whose similarities and dissimilarities could easily fall into a confused medley, a single structure of comparison has been adopted.

The cases are clearly contrasted. In only one (the UK) the site and designs were approved between 1976–79 and construction is now under way. Two others (the Netherlands and the FRG) actually got the plans approved, despite the inclusion by the Algerian suppliers of time-limit clauses that forced the pace of the negotiations. In neither the Netherlands nor the FRG, however, has construction work begun because, as a result of a change in their gas export policy in 1980, the Algerians cancelled the contracts. The Californian applications never got approval and were withdrawn by the utilities after about 15 years of investigating and negotiating with the government of Indonesia, concluding with a contract signed in 1975. There is reason for thinking that the Californian authorities would never have agreed to license a site, but the stated reason for letting the project drop was a change in the estimate of need for natural gas.

The UK planned to bring the gas from a domestic source; the Dutch and Germans were hoping to bring it from Algeria; and the Americans had supplies from Alaska and Indonesia in mind. None of these differences in sources seem to have played much part in the outcomes.

Decision-making analysis for these case histories has to be threaded through the differences between the more centralized, compact institutional structures in the UK and the Netherlands, and the loose federal constitutions in the FRG and the US. In the first two, the emphasis is more on discussion and consensus, while in the second two the procedures are frankly more adversarial. In the four cases, the points at which public participation was able to influence the decisions varied, as also did the points at which final risk analyses were called for and the issues on which they were focused. The local expectations of prosperity following the building of the plants and, of course, the extent to which those expectations were represented in the decision-making process, made a big difference. In Scotland everyone was convinced that this port and its hinterland, once a coalmining center but now very economically depressed, would get more work and income from the development. This was also the case, although to a lesser extent, in the Netherlands and the FRG, but much less so in prosperous California, where the manifold varieties of possible economic developments were much greater.

From the complexities, some conclusions emerge that are not unexpected but which are very interesting for all that. One strong impression is that the more the structure of the dialogue about risk is set in an adversarial institutional structure, the less likely is it ever to come to anything but a negative conclusion. Another fairly predictable impression is that the risk debate will come to a conclusion that accords with the negotiating strength or representation allowed to the party with the strongest views about the issue. These apparently homely insights are enormously important in the light of the two chapters on expert risk assessment. These provide a mind-boggling survey of different methods of measuring risks and the huge differences in the outcomes following small differences in the measures used. The experts disagree. But all the more does the apparently objective tool of risk analysis lend itself to capture by parties in the discussion. In the case studies the focus in the discussion shifts between whether there is any basic need to import LEG at all (a question in California and the Netherlands), whether the risks are acceptable (and what they are), and from this the identification of maximal possible bad outcomes, regardless of assigned probabilities of accident occurrences. There are huge degrees of freedom in analytical judgment that can push the measurement in any desired direction.

Apart from the history, the analysis of decision making cross-nationally, and information about discrepancies in the measures of risk, the book contains also some profound reflections on, for example, the implications of these methods (of democratic problem solving and compensation of losers) for distributive justice. Finally, within the same covers, an anthropologist makes reproaches to the policy analysts who have reduced cultural diversity to a single theoretical framework and who, in doing so, by imposing their own cultural viewpoint, have lost a lot of the explanations of what was actually happening in each country. The decision processes in the FRG, the Netherlands, and the US are unravelled again in a new contrast with the UK. A new pattern of forces emerges interacting with the legal and constitutional structures. This opens new vistas for analyzing cultural biases in the formulation of policies. Given the scope of its conception, the centrality of the issues involved, and the highly technical and theoretical level of analysis, this book must have profound implications for the basic dilemmas of Western social thought and for the workings of Western democracy.

Mary Douglas

June 1983

Preface

This book investigates the decision processes for siting liquefied energy gas (LEG) facilities in four countries: the Federal Republic of Germany, the Netherlands, the United Kingdom, and the United States. Different countries have specific institutional settings that influence how siting choices are made. A key thrust of the book is to explore the role of risk analysis, given a particular cultural and political structure, in siting decisions. It also suggests possibilities for distributing the total pie — i.e., costs and benefits — among individuals and groups affected by siting problems.

The introductory chapter sets the scene by providing synopses of the four case studies and highlights distinguishing features of the problem of siting LEG facilities. We then develop several reader's aids, including a multi-attribute, multi-party (MAMP) framework, to describe the sequential decision-making process associated with the siting of new technological facilities. These aids are applied to each of the four case studies in Chapters 3—6, thus enabling the reader to examine how agendas are established and how key parties interact with each other over time.

Chapter 7 compares different risk assessments prepared by analysts in each of the four countries, and reveals some rather wide discrepancies between estimates of the probability of accidents, even for the same facility. Since large-scale LEG facilities involve a relatively new technology, there are problems in measuring these risks. It is therefore difficult to determine the accuracy of risk analyses. In Chapter 8 these analyses are examined within the social and political context surrounding the siting debates. Particular emphasis is placed on the advantages and disadvantages of introducing quantitative risk analyses in political policy debates.

Chapter 9 focuses on ways to improve the decision process and the resulting outcomes. Particular attention is devoted to how policy analysis can aid the siting process. The chapter concludes by recommending four normative criteria for a desirable siting process, which are phrased in relative rather than absolute terms since each society will naturally determine its own appropriate targets.

During the course of the writing of this book, a cultural perspective on societal decisions involving risks to the public was developed at IIASA. In the Postscript this cultural approach is described and used to explain many of the anomalies found in the four case studies. This perspective both builds upon and, to some extent, challenges the framework developed in this book to describe political decision processes.

Howard Kunreuther
Joanne Linnerooth

April 1983

Acknowledgments

This book reports on a multidisciplinary effort to gain a better understanding of the decision processes associated with siting large-scale, potentially hazardous facilities in different cultural settings. It is primarily written for analysts and policy makers who have an interest in improving current procedures and learning about the experiences of other countries facing similar problems.

Although the process of standardizing and coordinating the various aspects of this study has rested primarily on our shoulders, the report is very much a group effort. The stage for the project was set by a IIASA Task Force Meeting on Liquefied Gas Facility Siting in September 1980, where participants in siting debates in five countries, representing government, industry, academia, and citizens' groups, offered us valuable information and advice. Following this meeting Michiel Schwarz and John Lathrop were very helpful in structuring the material to be collected for the case studies and providing guidelines for the interviews.

Christoph Mandl and John Lathrop were also responsible for the material dealing with the assessment and comparison of the LEG safety risks. Representatives of the interested parties in the four countries were interviewed by the case study writers: Hermann Atz (Federal Republic of Germany), Michiel Schwarz (the Netherlands), Sally Macgill (United Kingdom), and John Lathrop and Joanne Linnerooth (United States). We are grateful to all of the individuals who freely gave of their time and provided the case study writers with material and insights into the siting processes, and who later commented on preliminary drafts of the reports. During the year there were many additional persons who had a chance to review some of our material and provide us with helpful comments. A list of the individuals and their affiliations appears in a section at the end of the book, entitled "Consultants and Reviewers". The revised chapters reflect the many helpful comments we received from these people.

There are, of course, limits to recreating the dynamics of real-world decision making. To appreciate fully all of the nuances of the siting debates would have required team members to be flies on the wall and observe the many interactions between the interested parties. This activity was beyond the scope of the project, however!

Certain individuals deserve special mention because of their involvement in our research. Everyone on the project is greatly indebted to Eduard Löser of the IIASA library. He took a special interest in our research from its inception and provided us with published material related to all aspects of the project. Meredith Golden and Eryl Ley spent considerable time reviewing the case study summaries and other material in the document. Special thanks also go to Rhonda Starnes who superbly produced preliminary drafts of the manuscript singlehandedly and endured endless revisions with good humor and grace. Vivien Landauer helped coordinate the final phases of the project when she joined the Risk Group in early 1982.

The publication of this book would not have been possible without the hard work of Valerie Jones, who spent several months editing, and in many cases rewriting the manuscript. She did a superb job in coordinating the material of the many contributors from differing countries and disciplines. We also wish to thank Maria Bacher-Helm and Gertrude Maurer, with the valuable help of Hermann Atz, for their excellent work in translating the book from English into German.

During the past year we have had considerable input from IIASA researchers who have had an interest in the project. In particular, Nino Majone provided very helpful suggestions as to ways in which we could develop a framework for presenting the material in our four case studies. He also spent considerable time critiquing each of the studies and working individually with the case study authors. Michael Thompson provided us with numerous insights into cultural differences and styles between countries, and continually reminded us of the importance of castes and sects. James Vaupel commented extensively on the draft chapters of the report, and was particularly helpful in suggesting ways that the analyst can play a more constructive role in siting decision processes. Mark Pauly provided us with detailed comments and constructive suggestions on how prescriptive recommendations are related to the country's siting decision processes. Michael Stoto spent the summer of 1981 with the Risk Group at IIASA examining the role of experts in decision processes. In the same year Sally Blount prepared a paper on compensation systems that helped us to understand more fully their role in the policy process. Brian Wynne reviewed all of the case study material and provided us with numerous insights into the role of risk analysts in political decision processes. Participants at an IIASA Summer Study in June 1981 on Decision Processes and Institutional Aspects of Risk played a very useful role in critiquing our material, and especially in providing us with a broader perspective on risk-related problems.

The initial leadership of the project was under Craig Sinclair, who was at IIASA until December 1980. Howard Kunreuther coordinated the effort from January 1981. We want to thank Alec Lee, Chairman of the Management and Technology Area, Andrzej Wierzbicki, Chairman of the Systems and Decision Sciences Area, Roger Levien, former Director of IIASA, and C.S. Holling, Director of IIASA, who have encouraged the Risk Group in its efforts.

The entire project has been made possible by financial support from the Bundesministerium für Forschung und Technologie (BMFT) of the Federal Republic of Germany, supplemented by internal funds from IIASA. We are particularly indebted to Hans Seipel and Werner Salz, who were responsible for the program of which this study is a part; they have consistently provided us with encouragement and support.

Finally, a special note of thanks is due to our families, who were most supportive and understanding of the late hours that were inevitable at critical stages of the project.

HK, JL

Contents

Chapter 1

The Problem*

The Gasworks company has filed an application to locate a liquefied energy gas (LEG) terminal in the vicinity of Pietersdorf. Gasworks feels that this proposed project is economically justifiable from an investment point of view. It has also been encouraged to undertake such a project by the National Energy Board, because LEG promises to meet some of the country's future energy needs. The community of Pietersdorf stands to benefit from the plant insofar as it will provide additional tax revenues and future employment opportunities. However, some residents from the town are concerned about the impact of the facility on their future safety given the risk, even though relatively small, of a serious accident that might damage property as well as threaten lives. In addition, this group is concerned with the potential negative effect of large-scale technological facilities on the future quality of the environment and the lifestyle of their children and grandchildren. Environmentalist groups have voiced similar concerns and are opposed to the project.

The above scenario is typical of a wide class of problems where there are potential conflicts between different groups. Public concern over the siting of new technological facilities has been aroused in recent years through heated debates regarding the desirability of nuclear power as a source of energy. Proposals for new LEG terminals have created much less of a stir, but the principal issues are similar. In this book we explore the processes associated with LEG siting from a cross-national perspective; more specifically, interest is centered on the following questions:

- Why is the siting of LEG facilities a problem?
- Who is involved?
- What are the viewpoints of the different groups?
- How are the conflicts handled?
- Where do analysts and experts enter the process?
- How can analysts and experts play a more constructive role?
- What criteria should be utilized in making siting decisions?

These are discussed in the context of proposed LEG facilities in the Federal Republic of Germany, the Netherlands, the United Kingdom, and

*This chapter was written by Howard Kunreuther.

the United States. Prior to completing these four case studies, IIASA held a week-long workshop to discuss the similarities and differences between countries in siting LEG facilities (Kunreuther, Linnerooth, and Starnes 1982; henceforth abbreviated to KLS 1982). In various parts of this book we illustrate aspects of the decision process through selected comments from participants. At the end of the workshop, one of the participants, C.D.J. Cieraad of the TNO Center for Energy Studies in the Netherlands, had this to say:

> I think this has been a very informative meeting where the pitfalls of analysis, and the usefulness of it as well, have been explored. I think what we must include is that the use of analysis in politics must be very carefully looked at. (KLS 1982, p450)

This is the spirit in which this book is written.

SETTING THE STAGE

The oil crisis of 1973 and its subsequent sharp rise in prices shocked most of the world into recognizing that there is a tight interdependence between nations and that alternative sources of energy must be investigated. At the same time there was increasing concern with potentially harmful, long-term effects of new technological facilities on the environment, as well as the potential for catastrophic accidents. Hence, it is not a simple matter to recommend new sources of energy such as nuclear power or LEG in place of oil, even if these sources are cost effective. All aspects of human activity and concerns have to be considered in developing new energy strategies (Häfele 1981). It is against this background that our study of the siting of LEG facilities has been undertaken.

What is LEG?

Liquefied energy gas (LEG) comprises two substances: liquefied natural gas (LNG) and liquefied petroleum gas (LPG). The technology associated with producing and storing these sources of energy has been developed only relatively recently, but the number of new facilities has grown at a rapid rate. The principle upon which this technology is based involves chilling the gas until it becomes liquefied at a vastly reduced volume. For example, natural gas is chilled to $-161.5\,^{\circ}$C, at which point it becomes a liquid at about one-six hundredth of its volume at atmospheric pressure. Therefore, a tank of LNG contains 600 times as much energy as an equal-sized tank of natural gas.

The technology serves two principal purposes. First, it offers gas companies an opportunity to store its surplus gas during the warmer summer months by liquefying it. During the winter months when demand is relatively high, the LEG can then be regasified to satisfy peak demand. These so-called *peakshaving plants* can be linked to "satellite" stations in rural communities, which store LEG transported to them by road tankers.

Secondly, this technology offers a means by which natural gas produced in countries such as Algeria and Indonesia can be liquefied, shipped in specially constructed tankers, and received and stored at terminals hundreds or thousands of miles away. The LEG is then regasified and distributed to consumers mostly by pipeline, with the remainder carried by trucks or railcars. Figure 1.1 depicts the different elements associated with the transportation and storage process. The entire system (i.e., liquefaction facility, the LEG tankers, the receiving terminal, and the regasification terminal) costs more than $1 billion to construct.

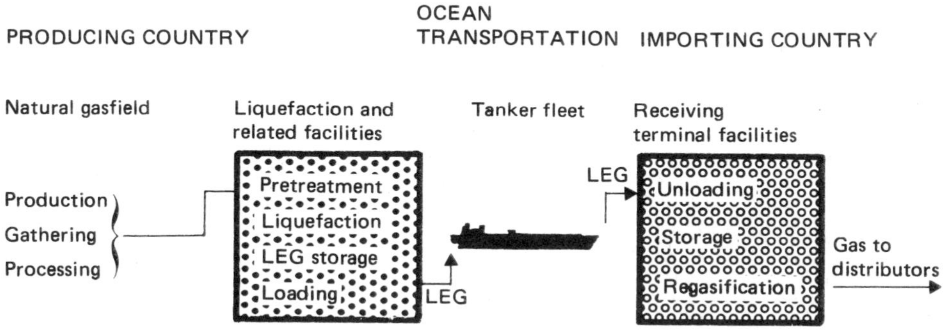

Figure 1.1. Major elements of a liquefied energy gas (LEG) project (adapted from Jensen Associates Ltd).

Of the four case studies reported in this book, three (the FRG, the Netherlands, and the US) involve the proposed construction of LNG import facilities while the project in the UK involves a facility for exporting LPG. As recently as 1977 the gas industry estimated that if all the announced international LNG import deals were to be put into operation by 1985, LNG could account for as much as 22% of American gas supplies, 23% of Western Europe's and 86% of Japan's (Daniels and Anderson 1977). Six countries import LNG and five more are considering constructing new facilities.[1] Figure 1.2 shows the location of existing liquefaction plants and storage facilities around the world. Davis (1979) points out, however, that in the late 1970s the potential of LNG as an important source of world energy diminished rapidly, for two principal economic reasons: a rapid rise in the price of LNG since 1973, and increasing concern over the reliability of supplies from countries such as Algeria and Indonesia.

How hazardous is LEG?

One of the most difficult problems associated with locating new technological facilities lies in assessing the safety risks to the surrounding population.

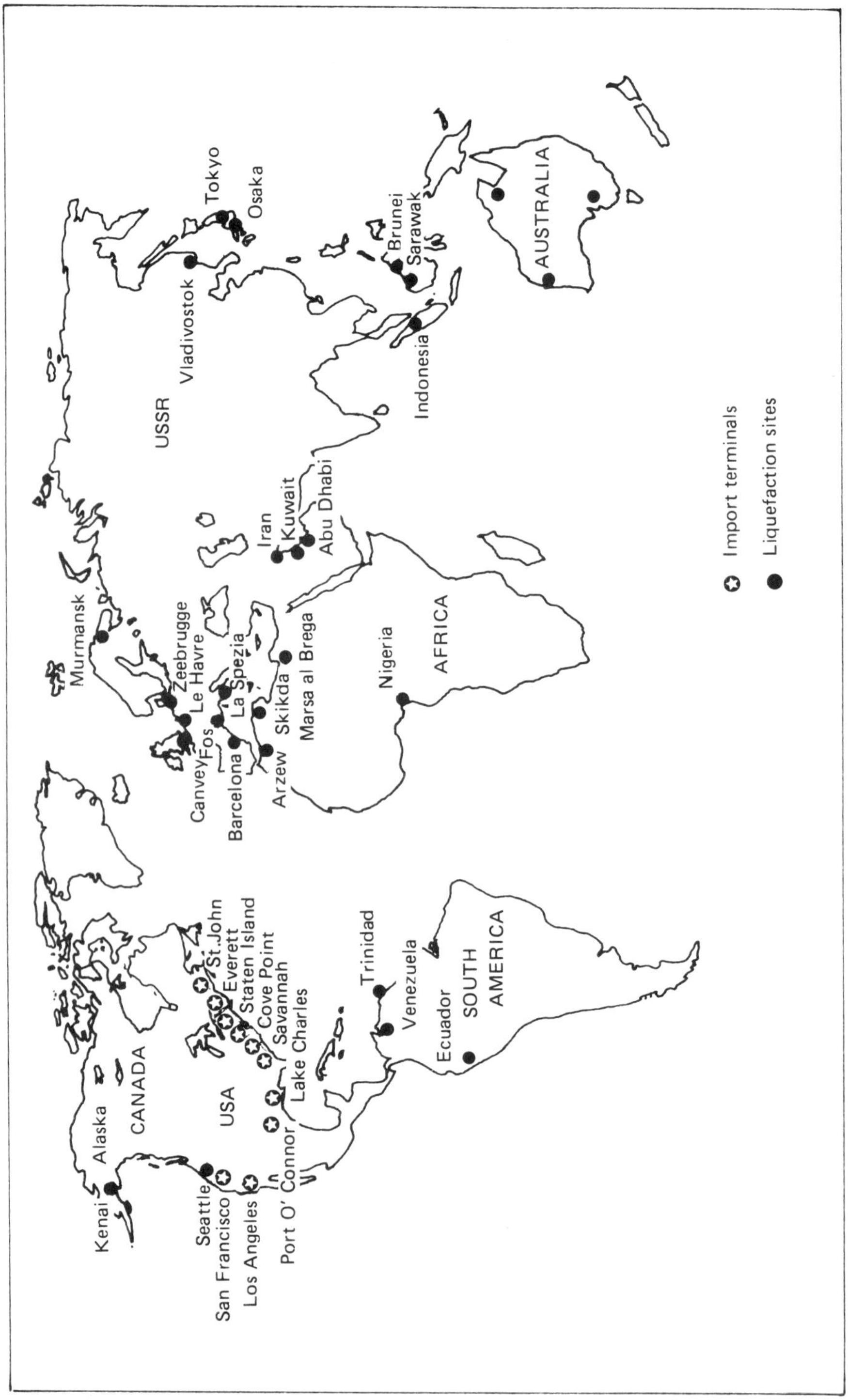

Figure 1.2. Map of the world showing the location of existing (1982) natural gas liquefaction plants and LEG storage facilities.

The development of LEG was scarred when the first storage tank, built in 1941 in Cleveland, Ohio, ruptured after three years of service, spilling the contents into adjacent streets and sewers. The liquid evaporated, the gas ignited and exploded, resulting in 120 deaths, 300 injuries, and approximately \$7 million in property damage. The plant was never rebuilt and set back the development of the technology for two decades. In the mid-1960s engineers and technicians were confident that the problems that had caused the tanks to fail had been corrected, so they pronounced LNG safe for future commercial development.[2]

To appreciate more fully the risks associated with an LEG facility, accident scenarios have been developed by analysts. Figure 1.3 illustrates the major components of such a scenario. The process begins with some initiating event, such as a vessel accident or a storage tank rupture, which results in an LEG spill. The LEG will immediately start to vaporize after a spill, producing a vapor cloud that can be carried downwind to a populated area. The mixture of vaporized LEG and air may ignite if it comes in contact with some ignition source, such as an electric spark. If the cloud does burst into flame, then there is a chance that fatalities may result from fire and explosions in confined spaces, and that materials will catch fire by radiant heat. Should the vapor cloud not be ignited within a given period of time then it will mix with a sufficient amount of air so that it becomes innocuous, and the cloud will disperse.

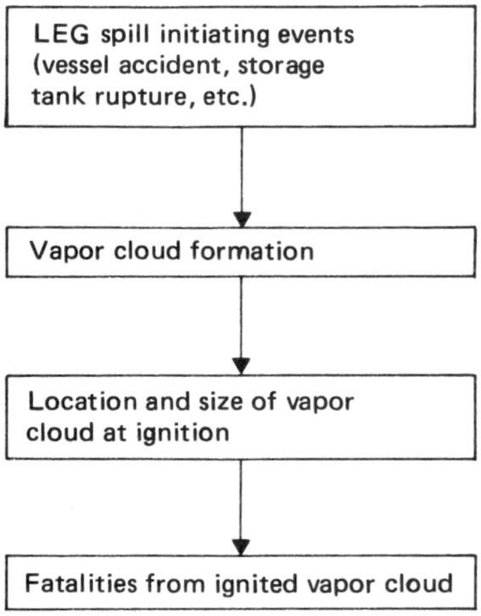

Fig. 1.3. Elements of an accident scenario (from Keeney 1980).

Scientific experts cannot agree on the risks associated with such accidents. What is the probability of an LEG spill? What are the chances that the resulting vapor cloud will ignite? How many lives will be lost and what will be the extent of injuries if such a sequence of events occurs? In Chapter 7 we document the wide discrepancies in these estimates due to different assumptions made by analysts on the causes and nature of these LEG-related risks. The widely different views on the potential hazards of LEG are important in understanding the political dynamics of the siting debates.

THE CASE STUDIES

The empirical basis of this book (Chapters 3—6) consists of case studies describing the decision processes and institutional arrangements associated with siting LEG facilities in four countries. The features of the processes were interpreted through retrospective interviews with key persons representing the relevant interested parties, as well as by using technical reports and documents supplied by these individuals. Interviews were conducted without any preconceived framework for analyzing the decision process; each case study writer simply obtained information as to how the parties perceived the different aspects of the siting process and in what ways their views changed over time. The roles of these groups were studied solely in the context of the specific LEG projects, rather than from larger agendas or perspectives on which parties may have based their actions. Below we present a brief synopsis of each of the four cases.

The FRG. In September 1972, two major gas companies founded a subsidiary for the purpose of building an LNG import/regasification facility in Wilhelmshaven, a port city on the North Sea coast, to handle LNG shipped from a liquefaction facility in Algeria. The terminal was expected to provide a significant proportion of the projected national energy supply for the FRG, as well as to aid regional industrial development in the Wilhelmshaven area. On the other side of the coin, certain parties contended that the proposed terminal would adversely impact the environment, and thus reduce tourism. There would also be increased risks to the population from shipping accidents that could occur at the berthing facilities in Jade Bay serving not only the LNG terminal, but also a proposed petrochemical plant and an existing oil refinery. Several risk studies were undertaken by interested parties and regulatory agencies, each of which looked at particular aspects of public safety. Conditional approval of the terminal was given in July 1979 by the Federal Minister of Transportation, who decided that the shipping risks were acceptable in view of the perceived economic benefits of the project. Because of problems with the Algerian gas delivery contract, however, no construction of the facility has taken place as of this date.

The Netherlands. In the early 1970s the semi-state-owned gas company Gasunie declared an interest in inporting LNG into the Netherlands, and started discussions with the Algerian company Sonatrach. The proposed site for the import terminal was at the port of Maasvlakte in the Rotterdam harbor area, since there was a confluence of interests among those concerned with energy policy and economic development in having a facility there. However, approval of a Maasvlakte site was endangered by opposition from environmentalist groups, so that Eemshaven was eventually selected over Rotterdam because the Dutch government felt a facility there would have a positive effect on the regional economy and employment. A significant factor that triggered the selection of Eemshaven was the feeling by some of the local parties involved in the approval procedures that the potential risks associated with the handling and reception of LNG at Maasvlakte were unacceptable. A government-sponsored risk assessment on LNG was used by all the major parties in the debate. Both the Eemshaven and the Maasvlakte locations were deemed acceptable by the government, but the former was approved in late 1978. Shortly afterwards, however, the LNG contract was canceled by Sonatrach so that, in the absence of alternative suppliers, further planning for an LNG import terminal in the Netherlands has been postponed.

The UK. In July 1976 Shell and Esso expressed a joint interest in locating a gas processing facility at Mossmorran and an export terminal at Braefoot Bay, Scotland, as part of a larger development program to process natural gas from the nearby Brent oil and gas field for export. Hence this project did not directly affect national energy supply, as was the case in the other three countries. The key points of controversy at a public inquiry were whether the potential economic benefits to the region would be sufficiently large to offset the perceived negative impact of the facility on the environment and, more importantly, the possible risk of a catastrophic accident. Amongst the individuals who potentially would benefit from the project were the residents of Cowdenbeath because of the jobs that would be created by the Mossmorran facility. The potential losers were the predominantly middle-class residents of Aberdour and Dalgety Bay, near the export terminal site at Braefoot Bay, who would be exposed to the risks of LEG storage and handling, and they therefore formed an action group to oppose the project. Independent risk studies were undertaken by the local authorities, the national government, and the Action Group. The site was eventually approved in August 1979 by the Secretary of State for Scotland, who stipulated that a detailed technical audit on the plant be undertaken prior to its commissioning. Today construction of the plant is well under way. It is expected to be in operation in the 1990s.

The USA. In September 1974, Western LNG Terminal Company, representing the interests of three gas distribution utilities, applied for approval of three sites on the California Coast: Point Conception, Oxnard, and Los Angeles, to receive LNG from Indonesia and Alaska. Over the next three years a number of interested parties at the federal, state, and local levels debated the desirability of each of these locations. By the summer of 1977, it became clear that none of these sites would be approved under existing procedures, so the utility companies and other business interests pressed for new state legislation to speed up the siting process. The result was the LNG Terminal Siting Act of 1977, which simplified the licensing process and stipulated certain population density requirements that would be acceptable. To help resolve the safety question, several risk studies were commissioned independently by interested parties, and these fueled the conflict between these groups. It appeared that Point Conception would be finally approved in October 1981 by the Federal Energy Regulatory Commission and the California Public Utilities Commission, since a recently discovered seismic fault was shown not to present an unacceptable risk. However, the question of whether California still needed an LNG terminal remained unresolved. The utilities have now announced their intention to withdraw their application for the present, but to keep open the option of pursuing the project further — possibly in the 1990s.

Similarities and contrasts

The political decision processes in each of the four countries reflect the different institutional structures within which groups debate controversial issues. Nonetheless there are certain common features that stimulated much of the controversy surrounding the development of LEG facilities. In the mid-1970s Western governments were searching for a more secure source of energy for the future, given the uncertainty of Middle East oil supplies. The energy companies felt that natural gas could fulfill this need and national governments were supportive of this development if the inherent dangers of such technologies could be shown to be acceptable.

If the choice of sites coincided with regional economic development needs, then the governments had an added incentive to favor a proposed plant. At the same time, the environmentalist movements were marshaling support for their concerns about the impact of large-scale technological development on the future quality of life and the risks to individuals residing near LEG facilities. Local action groups sprung up and voiced their concerns about the siting of these potentially hazardous facilities in their backyards.

In view of this conflict between national energy needs and safety/ environmental issues, it was inevitable that conflicts would arise between concerned individuals and groups. The following brief extracts from the four case studies highlight the difficulties that each country faced in dealing with these disputes.

In the FRG case, Hermann Atz considers the main failing to be an excessive amount of secrecy on the part of the government when considering plans for an LEG facility:

> In an atmosphere of growing societal concern about risk and the negative consequences of technological development it is advisable to reconsider the concept of strict isolation of decision making . . . without public debate. (Chapter 3, p60)

With respect to the Netherlands, Michiel Schwarz highlights the predominant influence of politics in determining the outcome of the debate:

> Although the government committed itself to an LNG terminal site, responding to an "energy policy" imperative, *political* considerations rather than safety concerns most strongly influenced the final outcome of the decision process. (Chapter 4, p92)

In the UK case study, Sally Macgill draws attention to the limitations of the British planning system in dealing with issues of risk:

> The risk studies used in the Mossmorran—Braefoot Bay decision process were part of a public debate that was unbalanced and inconclusive . . . . The debate was unbalanced in the sense that the emphasis given to various safety issues did not match the relative importance of those issues . . . . The debate was inconclusive given that the interested parties could not reach agreement on criteria on which to judge safety. (Chapter 5, p120—1)

In the US case study, John Lathrop and Joanne Linnerooth present the positive and negative features of the 1977 California LNG Terminal Siting Act, which gave a single state agency the mandate to grant a siting permit rather than requiring the utility to obtain approval from a number of local authorities:

> An LNG facility brings prospects of regional economic development and may increase the tax revenues of the municipality, but it also imposes risks on local residents. For this reason, a procedure requiring local approval may prove difficult. Yet, if citizen participation in decision procedures is important, any process that is less sensitive to local preferences may be undesirable. (Chapter 6, p144—5)

A look at the current status of the projects in all four countries reveals a radically different picture from the scene in the mid-1970s. Only the Mossmorran—Braefoot Bay export facility is under construction; the FRG and Netherlands have indefinitely postponed construction of their LEG terminals due to the uncertainty of gas supplies. In the US the utilities have concluded that LEG will not be a profitable future source of energy; hence it is unlikely that a terminal will be built at Point Conception in the immediate future even though the site has been declared seismically safe.

DISTINGUISHING FEATURES OF SITING PROBLEMS

There are several features of siting problems that make them a particularly interesting area for study.

Relevant interested parties

There are many different parties who have an interest in the final outcome, and each group has its own goals and objectives, which may change over

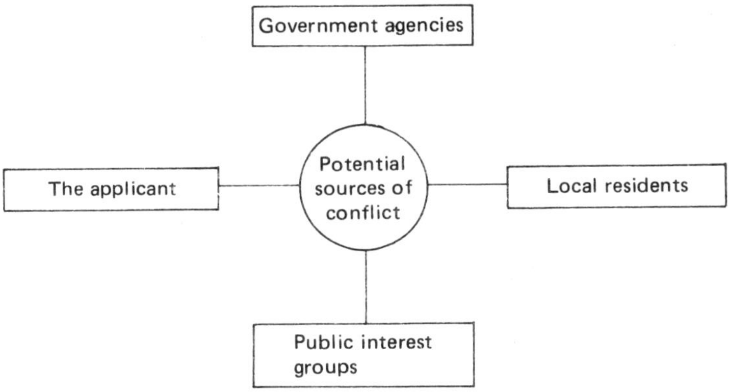

Figure 1.4. Relevant interested parties in a facility siting debate.

time. A closer look at the participants in LEG facility siting debates depicted in Figure 1.4 provides some insight into why conflicts are likely to result.

The applicant: Firms or developers who support the construction and operation of a facility have concluded that despite future uncertainties, the anticipated profits associated with the project will exceed the expected costs. Their analysis is likely to be based on the economics of operating a plant, the future purchase and selling price of the product, as well as the risk of a catastrophic accident.

Local residents: Residents in the vicinity of a proposed terminal are likely to have differing views of the situation. Those who own property on the land on which the plant is to be constructed may be concerned that a court will not award them a fair price for their property. Others may focus on the reduced property taxes, increased employment, or increased economic activity that a facility is likely to bring. A third group may be concerned with the potentially hazardous features of the facility. Hence some local residents may favor the project while others will oppose it.

Government agencies: Government agencies normally have well specified formal responsibilities in the siting process. Their roles, which are often defined by legislation, provide greater advantages to some interests than to others. The four case studies provide interesting contrasts with respect to the role of government agencies at the national, state or regional, and municipal levels, and show how the different stakeholders were affected by their actions.

Public interest groups: Recently there has been increased involvement by citizens' and environmentalist groups in siting debates. These organizations generally represent the interests and preferences of certain sections of the public. For example, the membership of the Sierra Club is concerned with the effects that the siting of any new technological facility will have on the environment. Local citizens' groups are generally concerned with how a proposed project may affect property values in the area and the safety of its residents. Those attracted to such organizations have strong, particular interests that dictate the agenda of the organization and influence the type of information that is collected and processed.

The following general objectives are at least implicitly considered by these different interested parties in judging the attractiveness of proposed alternatives:

- *Welfare objective*: the goal of improving the well-being of society; this is often referred to as the efficiency goal.
- *Distributional objective*: the goal of improving the well-being of specific individuals and groups; this is often referred to as the equity goal.

Certain interested parties such as the applicant, public interest groups, and residents are primarily concerned with their well-being and will favor a facility if they feel it promises to yield them net benefits over maintaining the status quo. Some government agencies may have a mandate to consider national interests, such as the impact of a project on the price and reliability of energy in the future. Those responsible for the environmental and safety aspects of a new facility will focus on the impact of the project on certain segments of the population. As we shall see in the four case studies, the institutional arrangements associated with LEG siting decisions influence the relative importance of welfare and distributional concerns in determining the final outcomes.

The contrast with, for example, a proposal to locate a shopping center in a community brings out some of the special features of the LEG siting problem. Those affected by the shopping center normally reside nearby so the project is likely to raise only local interest and concern, although this may still erupt into a heated debate. A proposed LEG facility will have a much wider positive impact since the potential beneficiaries cover a much wider region. Those who bear the costs will normally be locally concentrated, although national public interest groups may champion their cause. Hence the conflicts are likely to be between a large number of groups.

Louis Clarenburg of the Rijnmond Public Authority in the Netherlands highlighted the national conflicts between efficiency considerations and local equity concerns with respect to siting an LEG facility:

> I am sure that the majority of the population would be in favor of a guarantee of electricity. Only people living in the immediate surroundings of the site would probably be opposed. How to weigh the interests of the local population is the relevant question, and I don't feel that a referendum on the project resolves this question. It takes the responsibility away from the political level where it belongs. (KLS 1982, p232)

Sequential nature of the process

The political decision process associated with siting problems is dynamic and sequential in nature. Due to human limitations in processing and collecting information, as well as legislative and legal constraints, there are a number of subproblems that comprise the larger question of whether a project should be approved or not. The stakes of the different parties, as well as their relative importance, will vary over time depending on the particular subproblem being discussed. This feature highlights the importance of structuring the order in which questions are resolved. A different agenda for the same problem may lead to another final outcome because of the new clusters of interested parties who will interact with each other.

Exogenous events such as disasters help to structure the political agenda by calling attention to the dangers associated with a particular technology. A small database for judging the frequency of low-probability events increases the impact of these salient events on the decision-making process. Controlled laboratory experiments by psychologists have illustrated this type of estimation bias. For example, Lichtenstein and her colleagues (1978) have shown that individuals tend to overestimate the frequency of deaths from events that are easily recalled, such as natural or technological disasters. Their studies reveal that accidents are judged to cause as many deaths as diseases, whereas diseases actually take about 15 times as many lives. Tversky and Kahneman (1974) describe this bias as an availability heuristic whereby one judges the frequency of events by the ease with which they can be retrieved from memory.

Norbert Dall from the Alliance for Coastal Management in California feels that individuals have a difficult time dealing with probabilities and hence are greatly influenced by recent past experience. He illustrates this point with an example related to siting LNG facilities in communities in the US:

> . . . in places like Staten Island, Oxnard, and Santa Barbara, the local perception of risk has been formed by proximate, tangible and easily understood events that became etched into a deterministic appreciation of risk. Past events in the all-too-human hazardous industries suggest "it can happen here" and the puritanical strain in us recommends prudence. (KLS 1982, p146)

Since individuals have difficulty dealing with statistical measures of risk it is likely that they will fault the experts should a disaster occur. The result may be a distrust of technology, as evidenced by the reaction of the public to nuclear power after the Three Mile Island scare.

Pervading our analysis of the decision process is the presence of conflicts between the interested parties because each group has a different set of concerns that will influence the type of information it collects, and the uses it makes of specific events such as disasters or accidents. The resolution of these conflicts is likely to depend on the institutional structure of a particular country with respect to its allocation of responsibility to specific parties at different stages of the decision process.

Chapter 2 develops a framework for exploring the dynamics of siting processes in different countries. We have introduced several reader's aids to compare the four case studies: a graphical representation of key events in the siting process based on the program evaluation review technique (PERT); a party/concern matrix for depicting the main parties and their expressed concerns; and a multi-attribute, multi-party (MAMP) framework to help clarify the arguments the parties bring into the policy arena. The case studies provide descriptions of how conflicts were dealt with and resolved in the context of the different institutional structures that shaped the sequential political decision processes.

Disagreement on risks

Interested parties often disagree on the nature of the risks associated with new technologies such as LEG. Differences are created at even the most elemental level since the word "risk" has many interpretations. In this book we utilize the definition of risk provided by Fischhoff and his colleagues (1981b), that is, a threat to life or health as a potential consequence of a given activity. An activity is risky if it creates the possibility of these types of losses.

Weinberg (1972) was one of the first scientists to call attention to the estimation problems associated with low-probability events. He proposed the term "trans-science" to indicate that there is no practical basis for estimating the statistical chances and consequences of the occurrence of certain types of accidents although, epistemologically speaking, they are questions of fact. A similar point has been explicitly made in a report by the Council for Science and Society (1977), which indicated that the calculation of these numbers is highly uncertain. If interested parties provide alternative estimates of the uncertainties and consequences of these types of risks, it is particularly difficult to settle these differences in a scientific manner.

The problems associated with estimating risks has been highlighted because of a concern in siting debates as to whether a technology is acceptably safe.[3] At the IIASA LEG Task Force Meeting, Richard Mehta, chairman of a citizen's action group in Mossmorran, Scotland, suggested that some acceptable risk standard be made explicit:

> . . . there ought to at least be some generally acknowledged standard so that the public can judge what standard of acceptability was being implied. (KLS 1982, p196)

Harry Otway from ISPRA in Italy took issue with this point of view:

> I would warn Mr. Mehta against agreeing to a numerical acceptable risk criterion, because there is more to life than just not being dead — there are other things that matter besides physical risk. To what extent is this silly numbers game of "acceptable risk" really a surrogate for a debate about the legitimacy of insitutions? If the institutions were perceived as being legitimate, then risk should never emerge as an

> issue. Do discussions of acceptable risk beg the question of institutional legitimacy? (KLS 1982, p196)

Two surveys of acceptable risk problems have recently been made by Fischhoff *et al.* (1981b) and Vaupel (1982). Fischhoff *et al.* argue that no one solution to the problem is likely to be found because the alternative approaches have different strengths and limitations. The authors indicate that there are a number of uncertainties that make it extremely difficult to arrive at a magic number with which to guide risk decisions. In particular, they point out that there is uncertainty associated with defining particular decision problems, assessing the facts and values, and assessing the quality of the decisions that are made. In addition, both the general public and experts are fallible in their perceptions of risk and may exhibit systematic biases.

Vaupel points out that all of the acceptable risk approaches envisage a health, safety, or an environmental standard as being made by a single decision maker. Thus it is natural from these perspectives to imagine the decision being made by one "decision maker" who is given some facts and then decides what to do. Sometimes this "decision maker" is thought of as a person who has some discretion and can exercise judgment; sometimes the "decision maker" is, in effect, a formula or rule; sometimes the "decision maker" is a reification of the consensus of a group of individuals. In any case, the implicit image is that of a single decision rather than of a decision process. Hence, Vaupel argues, acceptable risk approaches may be misleading, inappropriate, and even unhelpful in some situations, like the siting of an LEG terminal, where the evolving trajectory of a policy is "the resultant of a complex process of interplay among many actors."

On a more general note, Wynne (1982) has suggested that we study risk perceptions not as a separate entity but rather as a part of the political and social organizational context in which decisions are made. This viewpoint is similar to that of Mary Douglas (1982), who claims that:

> Accepting risks is part of accepting organizations. The risk analysts and risk perception psychologists try to strip the idea of acceptable risk free of political adhesions, but the problems of risk perception are essentially political. Congresses and parliaments give away their rightful territory when they hand over such problems to risk experts. The public debates about risk are debates about politics. (p30)

In this book we take the viewpoint that risk is a political problem. Controversy frequently arises in siting debates because different stakeholders provide different estimates of the risk to reflect their views as to whether a particular technology should be part of society's portfolio. The four case studies presented in Chapters 3–6 delineate the roles of the major parties, the siting processes, and their attitudes toward the proposed LEG projects.

THE USE OF ANALYSIS

Chapters 7, 8, and 9 of the book explore and discuss the use of analysis as it affects the decision process for siting hazardous facilities. By analysis we mean a systematic set of approaches for dealing with a particular problem. Sometimes analysis is done in universities and research institutions; at other times it is done by consultants and advisors to interested parties.

Chapter 7 is devoted to a comparison of different analyses associated with the risk to life from LEG-related catastrophic accidents for the planned terminals in the FRG, the Netherlands, the UK, and the US. Since LEG is a relatively new technology there are problems in measuring these risks. The absence of a detailed statistical database has thus forced scientists either to use historical data or to develop physical models of the process. For example, the probability of an accident involving a vessel carrying LEG could be estimated by looking at the safety records of oil tankers and modifying these figures on the basis of anticipated differences in LEG tankers. Alternatively, ship collisions could be analyzed using an abstract model of shipping movements, taking into account traffic patterns for the harbor in question.

Risk analyses cannot be viewed in a vacuum but must be examined within the social and political context surrounding the siting decisions. In fact, the usefulness of analyses will be affected by the view that each society has towards its experts. In recent years the scientific community has lost its impartial image because technological uncertainty has allowed each scientist a wide range of interpretation regarding the risks associated with hazardous technologies. For example, with respect to nuclear power, scientists make judgments regarding the seriousness of health and environmental risk that correspond to their attitudes toward nuclear power (Nelkin 1981).

Chapter 8 compares how risk studies were incorporated in the decision procedures of the four case studies. Population safety was an important concern in all the siting debates, so that some type of risk analysis was required before each terminal could be approved. The national decision-making styles of each country determined how these analyses were actually used. Chapter 8 also investigates whether risk analyses were a useful input to the decision processes and why more comprehensive analyses were not pursued. Some suggestions are given as to how risk analysts might play a more productive role in setting policies involving the introduction of hazardous technologies.

Chapter 9 is concerned with how analysis can improve the siting process given the prevailing tensions between the welfare and distributional objectives. Our general view of how to proceed corresponds to the position taken by Randolph Deutsch, an attorney with the California Public Utilities Commission:

> The analysts should find out first what the real world is like, and then make an honest decision on where he thinks his analysis can be useful in the decision making process. I don't think it can be of use all the way across the board to decide all the societal issues in the siting process. (KLS 1982, p262)

Analysts can propose tools and approaches for facilitating the decision process itself, as well as the outcomes of party interactions. In focusing on the process side, attention is given to questions of *procedural rationality* — the way decisions are made. In addressing the outcome side, analysts are concerned with *substantive rationality* — the effect of different programs on the distribution of scarce resources among opposing groups.[4] We delineate three perspectives that analysts can take in improving the process and outcome side. Chapter 9 concludes by proposing a set of normative criteria for a desirable siting process.

A parallel approach, throughout the lifetime of this project, has focused exclusively on the cultural dimensions of the case studies, and its results are presented in the Postscript.

SUMMARY

This book is primarily a descriptive analysis of the decision processes and the use of risk analyses in siting new technological facilities. We are interested in the similarities and differences between four countries (the FRG, the Netherlands, the UK, and the US) in locating LEG terminals. By pointing out commonalities we indicate certain fundamental features of these siting problems that cut across cultures. By indicating differences we highlight the roles that different institutional arrangements and national styles play in structuring the decision processes and final outcomes. Based on these descriptive findings we suggest ways that analysts can play a more constructive role in the policy process.

NOTES

1. The countries currently importing LEG are France, Italy, Japan, Spain, the UK, and the US. Those who have considered investing in terminals are Belgium, Canada, Netherlands, the FRG and Sweden (Davis 1979, p11).

   An investigation of the Cleveland accident revealed that the tank had failed because it was constructed out of 3.5% nickel steel, which becomes brittle when it comes into contact with the extreme cold of LEG. All plants are now built with 9% nickel steel, aluminum, or concrete, and the storage tanks are surrounded by dikes capable of containing the contents of the tanks if a rupture occurs.

3. One of the earliest treatments on the subject of acceptable risk was by Starr (1969) who suggested that new technologies are acceptable only if they do not exceed the level of risk of existing technologies which provide similar benefits to society. This "revealed preference" philosophy has led interested parties to focus on acceptable risk levels that can be used to justify their positions.

4. See Simon (1978) for a more detailed discussion of these two types of rationality.

Chapter 2

The Framework*

Comparing four case studies presents an analytic challenge since each country has its own style, decision process, and institutional arrangements. This chapter presents a set of unifying concepts that are sufficiently broad to enable us to structure the siting procedure in each country using a common framework. Building on these concepts we develop a set of reader's aids that clarify the sequence of key events, the main concerns of the interested parties, and the arguments each stakeholder brings into the policy arena. Most importantly, these aids indicate how each siting problem was initially formulated and how it evolved over time.

UNIFYING CONCEPTS

Who has power?

Power is a major explanatory concept in political science studies of social choice. By power we mean the ability of an individual or group to influence the actions of others.[1] The concept can take several different guises. A party can have power by virtue of its *entitlement* or legal right in a specific situation (Calabresi and Melamed 1972). Individuals who own their own property have entitlement to the land and can determine whether they want to sell it to a potential buyer. Local governments also have entitlement to the land to the extent that they impose zoning restrictions, which constrain certain individual transactions. If a higher level of government sees a need for a certain piece of property, it may then employ special powers (such as eminent domain in the US) to obtain entitlement to this land.

Parties can obtain power through *standing*, by which we mean the right to participate in the decision process as specified by existing institutional

*This chapter was written by Howard Kunreuther and Joanne Linnerooth.

arrangements. For example, special interest groups have standing if their views can be officially voiced at public hearings as part of the siting process. If individuals have entitlement then they automatically have standing, but they can have standing without any entitlement.

Those parties who have *responsibility* for certain actions also have power. By responsibility we mean the formal duties specified by legislation as well as the concerns or mandates associated with a particular interest group. Government regulatory agencies are good examples of interested parties with formal responsibility; citizens' action groups and environmentalist organizations are examples of parties that feel they have a responsibility to their members to defend certain positions.

These three concepts of power can be illustrated in the context of the siting of LEG facilities. For example, in the FRG case study the municipality of Wilhelmshaven had *entitlement* to the land, which was zoned for industrial purposes. In the UK a local action group from Aberdour and Dalgety Bay was given *standing* by being allowed to present arguments against the proposed Mossmorran terminal at a public inquiry. In the Netherlands, the national cabinet had formal *responsibility* for selecting the terminal and for determining whether the safety risk was acceptable. In the US the California Coastal Commission had formal responsibility for protecting the California coastline and hence approving LNG projects until the passage of the LNG Terminal Siting Act, which transferred primary responsibility to the California Public Utilities Commission. Representatives of the Sierra Club reflected their concern and responsibility with environmental issues by opposing remote onshore siting in California in public hearings on LNG terminals in the state.

Bounded rationality

There is a growing recognition in the political science literature that decision makers are limited in their ability and desire to collect information on which to base their actions. They thus attempt to satisfice rather than optimize. One of the earliest descriptions of this bounded rational behavior in the context of societal decision making was by Lindblom (1959), who contended that the political process is one of incremental muddling rather than comprehensive choice. Instead of examining the full range of alternatives available, government agencies or politicians focus only on a limited set of options. They proceed incrementally by comparing the results of each new policy with old ones, thus drastically simplifying the decision-making process from the one implied by the classical rational model of choice (Braybrooke and Lindblom 1963).

Implicit in the concept of incremental decision making is the assumption that individuals and interested parties have a very limited amount of time available to deal with any particular problem. An excellent illustration of this feature is Aaron Wildavsky's (1964) analysis of the US budgetary process. Due to the complex structure and myriad sets of figures in the budget it is necessary for officials to employ simplified tools in making their choices. One of the principal ways budget officials justify their actions is to use the previous year's budget as a guide. In fact, as Wildavsky points out,

> Budgeting is incremental, not comprehensive .... Thus the men who make the budget are concerned with relatively small increments to an existing base. Their attention is focused on a small number of items over which the budgetary battle is fought. (p15)

The decision to build an LEG terminal at a particular location is also not made in a comprehensive manner in the sense that the alternatives to natural gas are scrutinized with full knowledge of the best available sites and the technical difficulties. A series of decisions are typically made, often beginning with approval "in principle" for the import or export project without full knowledge of the available sites or the technical details. This decision may partially lock the process into a certain course of action, and future policies are set on increasingly narrow aspects of the problem. For example, in California the question of whether California needed an LNG terminal was decided by the California legislature, which subsequently instructed the responsible authorities to identify an appropriate site.

Parties and issues

The literature on both organization theory and political science provide an important perspective on the societal decision-making process by stressing the role of multiple parties, each of which has its own goals and objectives. For example, March and Simon (1958) and Cyert and March (1963) view the organization as a coalition of parties, each imposing different demands on the system. The goals of the firm arise through a process of bargaining between potential coalition members. In a similar vein, Neustadt (1970), in his reflections on presidential power, points out that each of the interested parties in the government has its own interests and separate responsibilities. Policy emerges as a result of political bargaining between the actors.

One of the finest studies illustrating the importance of multiple parties in the decision-making process is Allison's (1971) analysis of the Cuban missile crisis. Of the three models he develops to explain the way policy is made, his government politics model (model III) comes closest to our view of the societal decision-making process. Allison points out that in decision-making situations there are many actors who are in the game as players, each focusing on multiple problems rather than single issues, and each having its own set of national, organization, and personal goals. The parties share power and have conflicting preferences. In order to determine how a policy emerges it is necessary to identify the various issues that are deemed important, to indicate what bargains and compromises emerge and "to convey some feel for the confusion" (Allison 1971, p146).

The sites that were selected for LEG terminals in the four countries studied were each a result of political bargaining, and in no case is a single interpretation of the recorded events adequate. One purpose of recording these events is to reveal the confusion with the hope that some insights will be gleaned, which will enable analysts to bring some clarity into a "muddled" world.

Conflicts and agendas

If there are a number of different actors in the societal decision-making game, then conflicts are likely to emerge. These conflicts can arise even if it appears that the interested parties agree on the overall objectives of a particular program. Simon (1979) provides an interesting example of the type of conflicts between groups in an early field study on administration of public recreational facilities. He found that there was general agreement between the school board and the city public works department on the objectives of the recreational program, but there was tension between the two parties on how the funds should be allocated between physical maintenance and play supervision. The reason for this was that the public works administrator viewed the playground as a physical facility, while the recreation administrator treated it as a social facility. In a similar manner interested parties may have differing views on an LEG facility, in which case there are likely to be conflicts and tensions between them on whether or not to approve the project.

One of the important questions that has been studied in recent years is how these potential conflicts are handled. Cyert and March (1963) hypothesize that sequential and decentralized decision making enables actions to be taken in many situations even if the parties concerned have incompatible goals. The importance of these features of the organizational decision-making process in a political context is highlighted by Herbert Simon (1967):

> Influence over the direction of attention of the political organs is a principal means for effecting action. The notion of power as a tug-of-war between alternatives yields to a notion of power as an influence on a sequential process in which actions must be generated as well as chosen, and in which attention is a scarce resource. (p108)

This characterization of the decision process is similar to the one formulated by Allison, who suggests that each party in the game faces an agenda with many deadlines, not all of which can be met. There is thus a need to set priorities. In other words, one needs to consider the nature of the agenda-setting process. As one would expect, those items placed on the legislative agenda become an important determinant of the final decisions that will be taken by society.

Cobb and Elder (1972) indicate that an important way in which an issue gets placed on the agenda is through some type of exogenous event that creates conflict. They illustrate this phenomenon using the example of the passage of the Federal Coal Mine and Safety Act of 1969, which was designed to reduce deaths from mine accidents and protect miners from "black-lung" disease. The legislation was triggered initially by a cave-in of a West Virginia coal mine that trapped and eventually killed 78 miners. This disaster caused the miners to strike, thus putting pressure on the state and federal government to react to the miners' concerns.

In another context, Holling (1981a) has pointed out how specific crises in the short term can lead to changes in policies with respect to environmental and ecological problems (e.g., the suppression of the spruce budworm after it had destroyed forests in Canada). Kunreuther and Lathrop (1981) describe with specific examples how exogeneous events triggered new coalitions and new legislation regarding LNG siting decisions in the US.

One reason for the importance of exogenous events, such as crises and disasters, in creating societal interest in a specific problem is that they indicate potential trouble in a salient way. Walker (1977) hypothesizes that accidents and disasters play an important role in setting the discretionary agenda of political bodies such as the US Congress. To support this conjecture, Walker presents empirical evidence on the passage of safety legislation in the US. Numerous examples of this process are also provided by Edward Lawless (1977) through a series of case histories of problems involving the impact of technology on society. He points out that frequently:

> . . . new information of an "alarming" nature is announced and is given rapid and widespread visibility by means of modern mass communications media. Almost overnight the case can become a subject of discussion and concern to much of the populace, and generate strong pressures to evaluate and remedy the problem as rapidly as possible. (p16)

In the case of decisions such as the siting of hazardous facilities, exogenous events such as an LEG explosion or an oil spill, may be sufficiently graphic and affect enough people to cause a reversal of earlier decisions, inject other alternatives into the process, or to change the relative strength of parties interested in the decision outcome. The mass media often focus on these specific events and in many cases exaggerate their importance.

Sequential processing of issues

Braybrooke (1974, 1978) has developed a concept of the political system which he views "as a machine or collection of machines for processing issues". In contrast to the static theory of collective choice based on the pioneering work of Arrow (1963) on social choice, Braybrooke views the decision-making process as sequential and constantly changing. At any point in time there is an issue or set of issues involving many different interested parties. Over time these issues may be resolved, disappear, or be transformed as new information or new alternatives emerge. In particular, new proposals may be constructed to reflect either the changes in preferences of the interested parties and/or a revised set of societal values. The importance of Braybrooke's work is that it enables one to decompose a problem into smaller subproblems by focusing on relevant issues. It thus captures the sequential decision-making process that characterizes individual and organizational problem solving (March 1978), as well as the public policy-making process (Gershuny 1981).

The setting of an agenda is likely to play an important role in determining the final outcome emerging from such a sequential decision process. Empirical evidence from the field, as well as from laboratory experiments (Plott and Levine 1978), indicates that the order in which specific subproblems are considered frequently leads to different outcomes for the same broader question.

We expect the same ordering effect for societal decision-making problems for two principal reasons. Once a decision has been made on a particular issue, this will then serve as a constraint for the next set of issues. If the order of the issues is reversed then there is likely to be a different set of choices to consider. Secondly, each issue involves a different set of interested parties, and each party uses its own set of data to support its cause. The timing of the release of this information may have an effect on later actions. For example, citizens' groups normally enter the scene with respect to siting problems only when a site close to their own community is being considered as a possible candidate. The order in which different locations are considered is thus likely to influence the final outcome of the debate.

During each stage of the sequential decision process there is a particular *arena* that serves as a focal point for the exchange of information. An arena is either a physical place like a courtroom, or a type of procedure like a public inquiry where interested parties either debate an issue or where an opinion is rendered. The types of arenas will differ between countries depending on the procedural system, as well as on the existing legislation that designates responsibility to each of the concerned groups.

READER'S AIDS

Here we present three "reader's aids" or analytical frameworks that provide a structure to the following case studies and that capture the concepts of the decision-making process described above. We introduce these aids with some hesitation, however, since analytical heuristics of any kind inevitably present a degree of artificiality to the process being described. Recognizing that paradigms might actually hinder understanding (Hirschman

1970), we nonetheless feel that a certain amount of structure is necessary for the reader to understand and compare the essentials of the four complex siting procedures. The three aids that we have developed for this purpose include a program evaluation review technique (PERT) diagram to illustrate the main events and decisions of each process; a party/concern matrix to illustrate the primary concerns of the main parties; and a multi-attribute, multi-party (MAMP) framework to illustrate the changing structure of the problems addressed and the interactions of the parties involved.

The PERT diagram

An important recent development in the analysis of decision-making problems over time is networking approaches. These techniques have been developed for analyzing "one of a kind" activities such as the expansion of a factory, the introduction of a new product, or a new computer system. We have turned to this literature in developing a standardized graphical format to structure the siting decisions in the four countries. More specifically, we have chosen to construct a diagram modeled on PERT.[2]

We incorporate only the schematic aspects of PERT in our treatment of the case studies. Figure 2.1 presents a hypothetical diagram in the form that will be displayed for each case study. The time line indicates the length of the process from when the project was proposed until a final decision was made (or to the present if the final outcome is still uncertain). In this example, the process was initiated in January 1970 and final approval of the project was given in December 1980. Here there are three

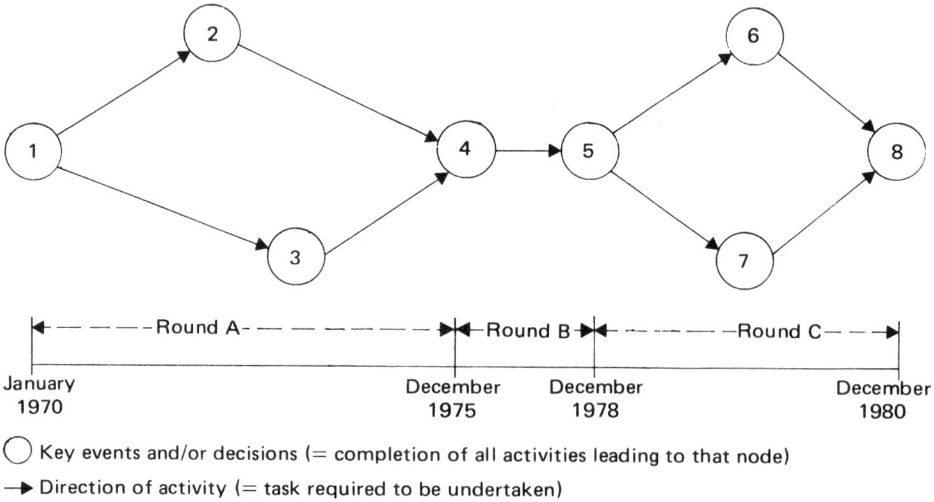

Figure 2.1. Hypothetical PERT diagram.

rounds, each characterized by a set of events. An event usually represents a distinct action taken by an actor or an interested party, although it could also represent an exogenous event such as a disaster. The arrowed lines connecting events indicate specific activities or tasks in the decision process. For example, event ④ occurs after the activities triggered by events ② and ③ are completed; event ⑤ triggers activities leading to events ⑥ and ⑦. This type of PERT diagram, as used in the case studies, provides a snapshot of the decision process and enables the reader to gain an overview of the activities, events, and groups participating in the debate.

The party/concern matrix

An important element of the case-study work has been to describe in detail the perspectives of each key party or actor entering the debate. These participants — or, more precisely, groups of participants — have certain concerns about the outcome of the decision. Their perspectives are often dominated by a particular dimension of the issue. In each case study, a table showing the major parties in the siting procedures and their principal concerns is presented. The table is intended to provide a general overview of the conflicting interests involved.

A list of the concerns expressed by the major parties in the four countries under study is shown in Table 2.1. This table is comprehensive only in the sense that it shows those concerns that selected persons representing the major parties revealed when interviewed; there is, of course, no guarantee that other "hidden" concerns (such as a psychological feeling against large technologies) exist. In some cases, then, these concerns may represent more the strategies of the parties than their "real" feelings about the project.

The MAMP framework

Although the PERT diagram and party/concern matrix present a snapshot of the key events and party concerns, they do not portray the dynamics of the process. For this purpose, we have developed the multi-attribute. multi-party (MAMP) framework, based upon Braybrooke's (1974) formulation of an "issue machine", which he developed to structure a case study concerning traffic congestion in London.

The anatomy of rounds. A central concept of this "issue machine" is the notion of *rounds* of political interaction to describe the evolution of public policies. A round is simply a convenient device to illustrate a change in the focus of discussions; that is, a change in the problems being addressed and a possible accompanying change in the parties involved. This new focus or formulation may be triggered by:

Table 2.1. Comprehensive list of actor/party concerns.

National	
Need	N_1
Import policy	N_2
Export policy	N_3
Economic benefits (taxes, royalties, balance of payments)	N_4
Economic costs	N_5
facility costs	N_{51}
harbor costs	N_{52}
channel costs	N_{53}
Image	N_6
Socioeconomic benefits	N_7
Energy policy	N_8
Regional	
Air quality	R_1
Industrial development	R_2
Economic costs	R_3
Support of smaller gas supply companies	R_4
Local	
Amenity and environment	L_1
land use	L_{11}
fishing, marine	L_{12}
tourism, recreation	L_{13}
local history	L_{14}
Economic benefits	L_2
taxes	L_{21}
jobs, short-term	L_{22}
jobs, long-term	L_{23}
jobs, indirect	L_{24}
spin-off industry	L_{25}
harbor activity and prestige	L_{26}
Local	
Economic costs	L_3
facility costs	L_{31}
harbor costs	L_{32}
channel costs	L_{33}
negative effects on fishing and tourism	L_{34}
Risks to population	L_4
shipping	L_{41}
gas storage	L_{42}
earthquakes	L_{43}
radio sparks	L_{44}
jetty operations	L_{45}
Risks to workers	L_5
Physical suitability of site	L_6
Ease and safety of shipping	L_7
Consumer	
Proximity of users to facility	C_1
Price of gas	C_2
Supply interruption risk	C_3
due to need for gas	C_{31}
due to source country	C_{32}
due to weather	C_{33}
due to accidents	C_{34}
due to earthquakes	C_{35}
Applicant	
Profit	A_1
revenues	A_{11}
costs, facility	A_{12}
costs, harbor	A_{13}
costs, channel	A_{14}
Control over sources	A_2
Image	A_3
Physical suitability of site	A_4

(*a*) a key decision made on some subset of the overall siting problem (e.g., a decision by the national government that a terminal is "needed" in view of national energy needs);

(*b*) a statement reached due to conflicts between parties; or

(*c*) a change in the context of discussions due to an unanticipated event, the entry of a new party, or new evidence brought into the debate.

Although the structuring of the process in this way has proved to be a useful heuristic, it must be recognized that these rounds are not unique. Indeed, there are many ways in which the divisions could be made. We have attempted to find those points where there appears to have been a changed agenda, or a central problem, that served as the focus of attention for the interested parties. For example, an important problem in California was whether there existed a *need* for imported natural gas. Three years into the approval process, legislation was passed by the California legislature declaring that this need existed and that a site for an LNG terminal should be identified by a certain date. This changed the political agenda and marked the beginning of a new round of political interactions.

While in some cases these breaks appear natural, we recognize that inevitably in some cases the natural flow of events have been forced into this analytical framework. Yet, in general, we have found that the cases fit easily and usefully into this structure.

Each round, in turn, is characterized by a general problem formulation, an initiating event or events, the interaction of the parties involved, and a decision (or "nondecision") concluding the round. These characteristics are briefly explained below.

Problem formulation: A critically important aspect of the political decision-making process is the way in which the problem is formulated. For instance, is the question before the relevant authorities "whether to choose site X or site Y", or is the question whether to approve the import of LNG in the context of national energy needs. Generally, we find that as the rounds progress, the number of open questions is reduced as each is sequentially resolved, sometimes bounding the discussions of the next round. An early round may be opened by industry submitting an application for a facility at its chosen site. In this case, the open questions may include whether the facility is needed and whether the chosen site is the best alternative; or, if these questions have been resolved (e.g., within national energy policy and a siting plan) the open question may instead be framed as whether the facility can be operated safely at that site.

Problems are formulated by a process where earlier decisions and the power of those with decision authority essentially frame the issues on the table. There is no pretense here that all the parties involved agree with this representation. Indeed, each party has its own definition of the problem, and ideas as to how it would frame the issue. For example, an opposition group may not agree that the facility is needed. Yet, if this question has been resolved in an earlier round, such as was the case in California, it may prove difficult, though not impossible, for the group to reopen the "need" discussion. It is a common characteristic of interactive problem solving that many (perhaps most) of the participants carry in their minds a distinct version of how they define "the problem". There is thus continuing competition between the parties to frame the problem on the political agenda; for example, will the hearings on site X allow debate on the question of whether the facility is needed? The "problem formulation" as recorded in the rounds of the case studies does not represent a consensus between the parties, but rather indicates the power of certain parties to define the issues to be addressed.

Initiating a round: A round of discussions may be initiated by a formal application or an informal request. Informal discussions may be initiated simply by a request for information on the part of one of the parties, or a request for preliminary discussions. The particular form of these initiating requests may further define or limit the bounds of the discussion.

The rounds recorded in the case studies generally commence when the question of siting an LEG terminal gains status as an issue on the political agenda. Internal discussions by the industry concerning a site would not constitute an issue on the political agenda; nor would discussions by the national government on energy policy if there is not a specific concern about an LEG terminal.

While the case study rounds begin with a formal request or application by one of the parties, much information, analysis, as well as politics, enter into the problem well before the explicit decision-making process is begun. Long-standing beliefs and attitudes constrain the problem formulation as it finally emerges for explicit consideration; they also constrain the alternatives that can be considered. To some extent, therefore, solutions are decided upon before they become the focus of discussion, analysis, and debate (Lindblom 1982, personal communication). These "pre-rounds" of the process are unfortunately difficult to elicit since there is little formal documentation, and parties often have conflicting interpretations of the activities. To the extent possible, however, these early interchanges are discussed in the case studies.

Interaction phase: To understand a particular pattern of institutional choice it is necessary to describe the actors, their respective responsibilities, interactions with one another at different stages of the process, and the information available to them. In a formal sense, a party's evaluation of a specific siting alternative is based on its estimation of the levels and values of each effect or attribute resulting from that alternative, and the relative importance given to each attribute. Another party might have different estimates of the effects of an alternative based on conflicting information, its own analysis of costs and benefits resulting from those effects, or a different ranking of the importance of the attributes.

Another important feature of the decision process is that the value of an attribute can change over time because of the introduction of new information. For example, if a report provides new insight into the seismic risk associated with a particular site, this may cause a change in the perception of this attribute by one or more of the parties involved. That change may take the form of a different estimate of the level of the attribute for that site, or perhaps a different weight may be given to the relative importance of the attribute.

The outcome of the political debate, to a large extent, results from some combination of the following: the formal responsibilities of the parties involved; the attention given to the issues in light of limited financial resources and time; the way in which the problems on the political agenda are framed; and the exogenous events that may change the problem and/or the parties. The interaction phase can be thought of as the formal and informal communication between the parties influencing the decision outcome. Wynne (1982) feels that it is futile to assume that parties interact as "rational" actors in the sense of actively and openly pursuing clearly defined objectives. Defensive behavior, that is, the avoidance of problems and dangers, may be as rational in these situations as goal-directed behavior.

In addition, Majone (1979) points out that organizational behavior is usually not directed at problem solving in any rational sense, but rather at serving the longer-term interests of the organization or institution. In the public domain, as opposed to the market, decisions must be justified using seemingly objective arguments. Therefore the points made by the parties are important insofar as they reveal the complex strategies and counter-strategies of those in the policy game.

The interactions between the parties, as recorded in the rounds of the case studies, are represented by the *main arguments* each party brings to the debate in support of or in rejection of each of the alternatives at hand. These arguments may relate to only one or two attributes. In some cases the attribute may be selected to maximize the effectiveness of the argument rather than reflecting an actual concern of the party. For example, a party opposed to a site because of its concern for environmental quality may present an argument using seismic risk as the main reason for rejecting the site. The argument thus reflects a *strategy* on the part of the actor in support of or in opposition to the proposal. This strategy may be based on hidden agendas that are never revealed by overt actions or statements. By studying the strategy of the actors, one could uncover underlying motives and desires. This is important in understanding the interpretation and use of scientific evidence, including risk analyses.

The interactions between the various and often multifold participants is only partly face-to-face, as in a committee, task force, or hearing. Most of the dialogue that bears on the final resolution of the policy question are distant interactions. Moreover, they sometimes occur without individuals being directly aware of the actions of others. The interaction phase provides useful insights into the process. Parties often come into the debate with firm preferences. The interaction phase brings out their arguments, and perceptions, and reveals many changes in their positions on an issue. The stability of the system can, at least partially, be judged by the degree to which the actors and arguments remain the same after each successive round.

Concluding a round: A round is concluded by a decision, a stalemate, a change in information (changing the focus of the debate and hence initiating a new round), or an exogenous event (e.g., a disaster) aborting the discussions and requiring a new round of inquiry. Each decision, in turn, can be described by the trade-offs implicit in the choices made. These trade-offs may not be explicitly recognized by the decision makers, nor explicitly analyzed in the process of making the decision.

The conclusion of a round can take one of two forms. If there is a feasible solution that is agreed upon by those with decision responsibility, the process ends. However, if one or more parties is dissatisfied with the situation at the end of the round, and has recourse to other channels, or if the round ends in a request for further action, a new problem is formulated for the next round. This sequence is repeated for another set of alternatives, interested parties, etc.

SUMMARY

In summary, the societal decision-making process is one where there are a number of interested parties each of which has its own goals and objectives. Each group also has its own set of information that is used to defend specific recommendations. The decision-making process is frequently sequential and decentralized, and many items compete for the limited time and attention of the people involved. The process of agenda setting is an important element in understanding why certain problems are considered important while others are ignored. Recent empirical studies have stressed the importance of exogenous events in having items placed on the formal agenda. Political decision making is likely to follow a sequential process whereby new issues emerge through the resolution of previous issues, changes in party preferences, and/or social norms.

Three aids have been developed to structure the political siting processes of the four countries: a PERT diagram to describe the main events; a party/concern matrix to list the concerns of the major parties; and a MAMP framework to describe the evolving problem formulations and arguments of the participating parties. The case studies themselves were developed after extensive interviews with key individuals involved in the siting processes of the countries concerned. Each interviewee was asked to describe the process and problems of siting an LEG facility from his or her own perspective.

Only after the individual perspectives of each of the major participants or observers were elicited and recorded in full-scale case studies did we develop the reader's aids. These aids are not intended as analytical tools. Rather, their purpose is to help trace and structure the events and party participation of four complex siting procedures in a way that lends itself to making descriptive comparisons. Analysis and interpretation of this material, such as questioning the usefulness of scientific expertise in political siting issues, is only possible with a firm understanding of the people and institutions involved, as well as of the procedural constraints.

In the following case studies the people, institutions, and political styles of the four countries in selecting and approving LEG terminal sites will be described using these reader's aids. Chapters 8 and 9 address some of the analytical questions raised, based on this descriptive material, coupled with the use of risk analyses in the four countries.

NOTES

1. This definition of power is used by Cobb and Elder (1972), who elaborate on its importance in the context of agenda building.
2. The PERT approach originated in 1958 as part of a research project designed to evaluate the progress of the US Navy's *Polaris* missile program (Emory and Niland 1968).

Chapter 3

The FRG: Ripples at Wilhelmshaven*

This chapter deals with the siting and approval process for a liquefied natural gas (LNG) import facility at Wilhelmshaven in the Federal Republic of Germany. The major aspects of the political decision-making process are summarized, focusing on the role of technical analyses of public safety risk in this decision. The most remarkable feature of this process was that despite the novelty of the LNG technology in the FRG, it deviated very little from established industrial siting and approval procedures. Public interest in the project and concerns about its acceptability did not rise above a relatively low level. At some point, however, unexpected difficulties related to the question of safety risk seemed to threaten the approval of the terminal, but these were eventually overcome by the federal government in a rather elegant way, leaving little more than ripples on the surface.

SETTING THE STAGE

The Basic Law (*Grundgesetz*) of 1949 established the FRG as a federation of what are currently ten autonomous states or regions, called *Länder* (excluding West Berlin, which has special status). As a rule, the federal government establishes laws and regulations that are implemented by the *Länder* (for a more detailed description see Southern 1979). There are five administrative levels: federal (*Bund*), state (*Land*), district (*Regierungsbezirk*), county (*Kreis*), and municipality (*Gemeinde*). The counties and municipalities act in some matters as agents of the states, but at the same time they are also responsible for executing the local autonomous legislation (*kommunale Selbstverwaltung*). At the two lowest levels an

*This chapter was written by Hermann Atz, based on a more comprehensive case study (Atz 1982). The author worked closely with Joanne Linnerooth in integrating the case study material within the MAMP framework.

elected council oversees and also usually appoints the head of administration. At the district level, no elected body exists, but the district government is responsible for a variety of administrative duties. It is the principal executive institution within the state, and may act as an agency of the state government in matters delegated to it.

Political decision making in the FRG cannot be fully understood unless one takes into account the specific separation of powers between the legislature and the executive as laid down in the constitution and perceived by the citizens. In contrast to the US system, for example, public administration is regarded as a branch of law, rather than as part of the political system proper. Because administrative bodies are not directly controlled by the elected bodies, the administrative law and its courts play a particularly important role in protecting citizens against decisions that they perceive to affect their granted rights. Consequently, citizens are quite capable of dealing with administrative procedures, whereas most of them exhibit passive attitudes towards the policy formulation process (Reichel 1981).

Because of this complicated distribution of responsibilities, the strict separation of powers, the important role of civil servants in policy making, together with the relatively weak position of individual political parties, a tendency towards consensus has prevailed in German politics (see Scharpf *et al.* 1976). Yet, over the last 15 years, a growing number of environmentalist and citizens' action groups of all kinds have entered the political arena, and in many cases siting decisions on large-scale technological installations (such as nuclear power plants) have been the focus of strong protest (see Murphy *et al.* 1979, Guggenberger 1980, Kitschelt 1980). Despite the extension of formal public participation, however, the opportunity for interest groups to influence particular decisions through political demands has remained rather small, although in some cases litigation has proved an effective means of stopping, or at least of delaying, certain industrial projects and technological policies.

Wilhelmshaven, established as a Prussian naval port in the nineteenth century, is a city of about 100 000 inhabitants on the German North Sea coast to the west of Bremerhaven (see Figure 3.1). In spite of its location on Jade Bay, which provides naturally good shipping conditions, Wilhelmshaven's industrial base has remained relatively weak, partly because of the city's dependence on military installations did not encourage such development, and partly because of other factors such as the great distance from the consumer markets in the Hannover area, Lower Saxony's economic and administrative center. The coastal zone near the city of Wilhelmshaven is one of the least developed parts of Lower Saxony, with a high rate of unemployment.

To help alleviate these economic difficulties Wilhelmshaven has been designated an industrial center in regional development plans. To attract industry, the shipping channel was deepened and a large area of land

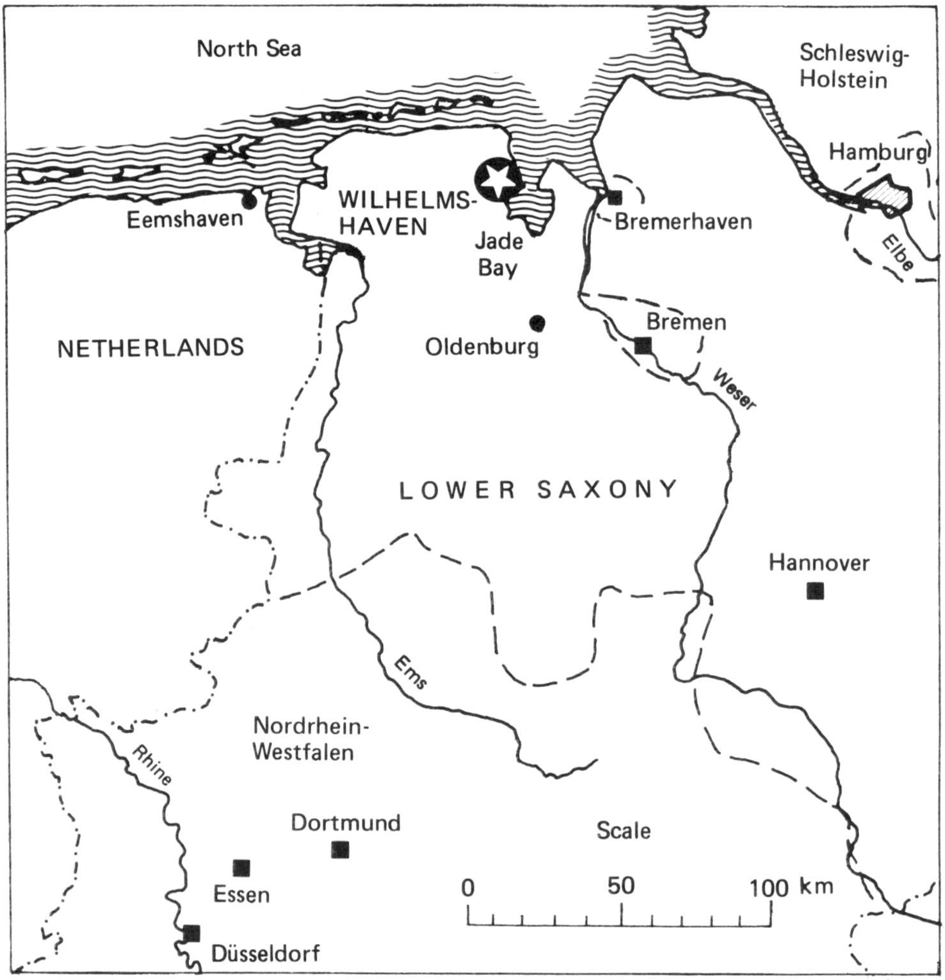

Figure 3.1. Northern part of the Federal Republic of Germany.

reclaimed from the sea. The state of Lower Saxony has been deeply involved with these activities, and the LNG terminal was discussed in this context in order to make use of the infrastructure created to promote industrial developmnet.

The site selected for the LNG terminal lies in the north of this reclaimed land near Wilhelmshaven, on the border of the adjacent municipality of Wangerland (see Figure 3.2), which has a population of approximately 10 000. The village of Hooksiel, which belongs to this municipality, is less than 2 km from the terminal plot, and a recreational area is only a few hundred meters away. This recreational zone was intended to compensate for the loss of Hooksiel's boating harbor and for the inconvenience resulting from the closeness of industry. In fact, the proximity of Hooksiel

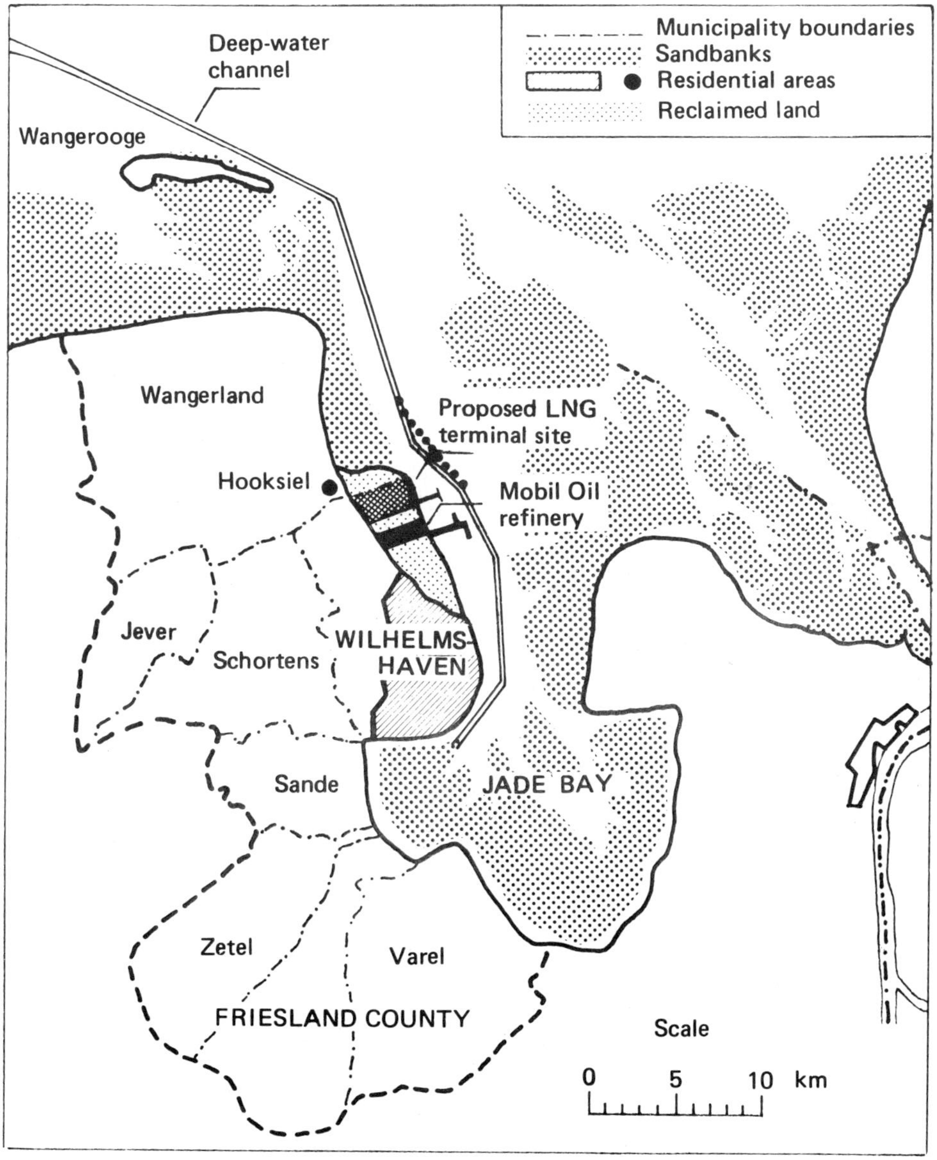

Figure 3.2. County of Friesland and Jade Bay. The dotted line shows the proposed change in the configuration of the deep-water shipping channel required by the WSB in the plan specification.

and the unfortunate location of the recreational zone became important points of conflict during the procedures relating to the LNG terminal (see Figure 3.3).

The project under consideration involved the construction of an LNG import terminal at Wilhelmshaven to store natural gas shipped from a

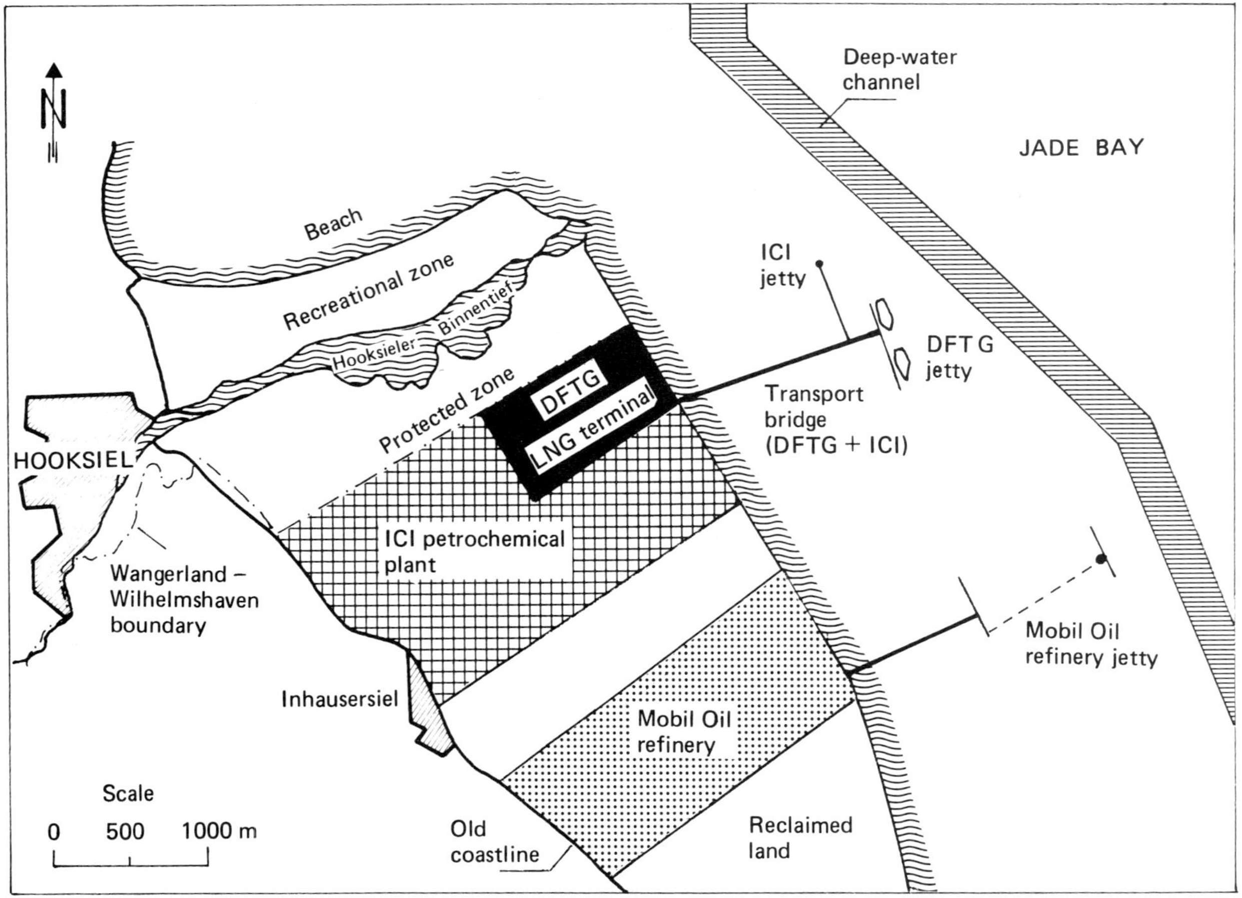

Figure 3.3. Detailed plan of the industrial zone at Wilhelmshaven.

liquefaction facility in Algeria in special tankers, each carrying 125 000 m^3 of LNG. From Wilhelmshaven the gas would be distributed to consumers either by pipeline (after regasification) or by smaller LNG tankers. The planned terminal consisted of a berth to load or unload one small and two larger LNG tankers at the same time, a closed transfer system to the facilities on land, storage tanks (originally four, later six) for approximately 500 000 m^3 of LNG, and a regasification plant.[1] The facility was designed to transfer 12 billion m^3 of natural gas per year (equivalent at normal temperature and pressure), corresponding to one-fifth of the current (1980) level of natural gas consumption in the FRG. As described below, the LNG project became closely bound up with plans to build a petrochemical plant handling hazardous chemicals (ethylene, vinylchloride, and soda-lye) on an adjacent site. Both facilities were designed to use a common transport bridge (elevated pipelines for LNG and liquid chemicals, and a road) connecting the two separate jetties.

Because of the sandbanks on Jade Bay, the planned jetties would have to be at least 1500 m long, extending to within 500 m of the deep-water channel. At this point there is a fairly sharp bend in the channel, and so a seriously debated concern was the risk of a vessel deviating from its course along the channel and colliding with an LNG tanker moored at the jetty or with the transfer facilities.

Different technical features of the LNG project determined which approval procedures would be necessary (see Steiger and Kimminich 1976). The most important were the following:

(1) The land-based facilities would be licensed, as is the case with other major industrial projects, by a state administrative agency.
(2) Since Jade Bay is within the three-mile coastal zone, it is considered to be part of the inland waterway system, under the jurisdiction of the Federal Waterways and Shipping Administration. A special licensing procedure is necessary for any construction in a federal waterway.
(3) Plans for a land-based facility can only be licensed if they comply with town and country planning regulations. The adjustment of the relevant zoning plans to suit the particular development (a responsibility of the municipalities) was therefore a precondition for the approval of the proposed terminal.

SEQUENCE OF EVENTS

After briefly introducing the principal parties, we will structure the main events with the aid of a PERT diagram, as described in Chapter 2. This discussion presents an outline of the main decisions and events in the siting process, but does not take account of internal decision making within each interested party.

The main parties that played a dominant role in the decision-making

Table 3.1. Key parties in the LNG decision process in the FRG.

Applicant	
Ruhrgas, Gelsenberg (DFTG)	Two important private energy supply companies and their subsidiary company, DFTG, applied for approval of the LNG terminal. Ruhrgas is specialized in the gas market, and provides two-thirds of the supply.
National government	
Federal Minister of Transportation (FMT)	Highest authority of the Federal Waterways and Shipping administration, responsible for maintaining and developing waterways and regulating shipping in them. It was in this last capacity that the FMT was involved in the LNG decision process.
Water and Shipping Board North-West (WSB)	One of the six federal waterway authorities at the intermediate level, in charge of licensing the proposed harbor facilities of the joint DFTG/ICI project in accordance with the Federal Waterway Law by a plan specification procedure.
State government	
Lower Saxony Ministry of Economic Affairs and Transportation (ME&T)	Executes regional development programs for the state (*Land*) government; in particular, it grants subsidies and advises interested enterprises in their siting decisions.
District Government of Weser-Ems (DGWE)	Involved in two different functions in the LNG siting process: responsible for granting a license for the land-based facilities according to the Federal Immission Control Law; and acted as an agent of Lower Saxony in planning a part of the harbor facilities, namely the transport bridge to be used jointly by DFTG and ICI. In this latter capacity, DGWE also filed an application with WSB for a plan specification procedure.
Local authorities	
Wilhelmshaven	Host municipality and county of the LNG terminal, comprising an elected council and administrative body headed by an appointed director. While the city council has to approve all main decisions, most of the actual work is done within the administration. Because of local autonomy with respect to zoning and development planning, compliance of the municipality is crucial for any industrial project.
Wangerland	The municipality bordering on the terminal site in the north of Wilhelmshaven. Since some Wangerland citizens would be more directly affected than the majority of residents of Wilhelmshaven, this municipality was naturally an interested party looking very critically at the project.
County of Friesland	Encompasses eight municipalities around Wilhelmshaven, excluding the city. The county had only limited formal duties in the decision process, but it strongly supported the interests of Wangerland, a municipality within its jurisdiction.

Table 3.1. continued

Other interested parties	
Hooksiel Citizens' Group (HCG)	The most effective action group intervening in the LNG siting debate, formed by members of 17 associations and clubs based in the village of Hooksiel. The group attempted to represent the interests of the residents in a quasi-official way, but its influence depended mainly on a few very active persons.

process are listed in Table 3.1. Other interested parties were either of negligible influence, or their positions were represented by other groups listed in the table.[2]

The PERT diagram showing the main events leading to the approval in principle of an LNG terminal at Wilhelmshaven in 1979 is shown in Figure 3.4. The circled numbers in the diagram and in the text indicate the decision events that can be construed as dividing the process into four partially overlapping rounds. Rounds A and B embraced the siting decision proper, outline planning of the facilities, and the general agreement between three of the parties — Ruhrgas and Gelsenberg, Lower Saxony, and Wilhelmshaven — on the desirability of the project. The initiation of the formal approval and licensing procedures is taken as the starting point of round C, in which most of the public debate on the LNG terminal took place. Round D dealt with the question of shipping hazards, which emerged as the most critical from the licensing procedures.

Rounds A and B. In the late 1960s several gas companies in Europe began considering the possibility of importing LNG from Algeria, at that time a major exporting country. After completing an internal screening process, Ruhrgas and Gelsenberg decided that they would prefer a possibly more expensive domestic site to other potentially less costly sites in Belgium, France, or the Netherlands, and this was justified by the perceived lower risk of supply interruption. In 1972 the two companies founded a subsidiary firm based in Wilhelmshaven to build and operate an LNG terminal, the Deutsche Flüssigerdgas Terminal Gesellschaft mbH (DFTG). ①

Subsequent consultations with the Lower Saxony Ministry of Economic Affairs and Transportation (ME&T)[3] and with local authorities confirmed DFTG's view that Wilmhelmshaven was particularly appropriate for an LNG terminal site. ③ In these talks the ME&T acted not only as the state authority concerned with regional economic development in general, but also represented the state of Lower Saxony as the owner of the reclaimed land on Jade Bay. The choice of Wilhelmshaven was agreed in principle in a preliminary contract between Ruhrgas and Gelsenberg, DFTG, and Lower Saxony in November 1973. ② These parties, and the Wilhelmshaven authorities, then engaged in long and tough negotiations concerning

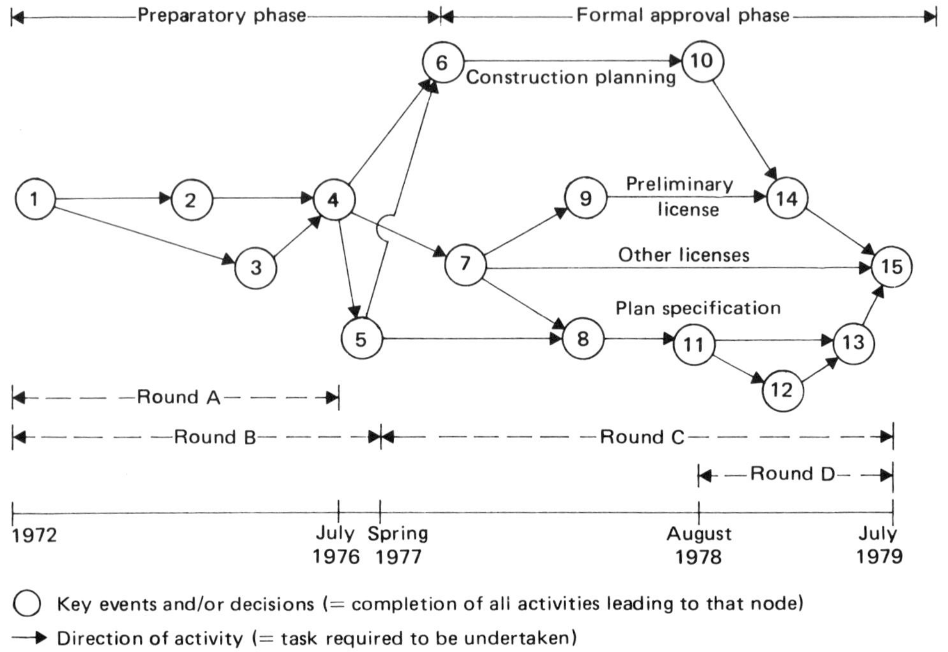

Figure 3.4. PERT diagram for the LNG decision process in the FRG.

Key events and/or decisions

(1) Ruhrgas and Gelsenberg decide in favor of a domestic site for the LNG terminal and set up the subsidiary DFTG (1972).

(2) Gas companies, DFTG, and Lower Saxony sign preliminary contract (Nov. 1973).

(3) DFTG, after consultations with public authorities, selects from possible sites at Wilhelmshaven (June 1974).

(4) A settlement contract is agreed between DFTG, Ruhrgas, Gelsenberg, Lower Saxony, and Wilhelmshaven (July 1976).

(5) DFTG, ICI, and Lower Saxony agree on joint harbor facilities for the petrochemical plant and the LNG terminal (spring 1977).

(6) Wilhelmshaven initiates the construction plan approval procedures for the terminal site (May 1977).

(7) A delivery contract between a Dutch—German consortium and Sonatrach of Algeria is signed (June 1977).

(8) DFTG, followed by ICI and Lower Saxony, files application for the harbor facilities with the WSB (September 1977 to February 1978).

(9) DFTG files application for land-based facilities with the district government of Weser-Ems (September 1977).

(10) Wilhelmshaven City Council approves the construction plan after overcoming objections raised by interest groups (July 1978).

(11) The Federal Minister of Transportation (FMT) becomes directly involved in the approval procedure concerning harbor facilities (July 1978).

(12) The FMT evaluates the risks related to the LNG terminal and tanker traffic on Jade Bay, and finds them acceptable subject to certain conditions (March 1979).

(13) WSB approves the joint harbor facilities (March to July 1979).

(14) The DGWE grants its preliminary license for land-based LNG facilities (July 1979).

(15) Termination of approval procedure for construction of the LNG terminal (July 1979).

necessary preconditions for approval of the project, the outcome of which was the signing of a "settlement contract" in July 1976.[4] ④

Simultaneously, at the request of the ME&T, the local authorities affected by the LNG project, the regulatory agencies, and DFTG considered its feasibility from different perspectives, such as impact on the regional economy, environmental effects, and occupational and public safety. They also considered suitable locations within the industrial zone at Wilhelmshaven and agreed on criteria relevant in selecting the actual site (see Table 3.3, round B). In 1974, DFTG decided in favor of one of the two plots offered by Lower Saxony, namely the northernmost zone. Legally, the sale of this piece of land was finalized in the settlement contract and was later approved by the parliament of Lower Saxony.

Following the negotiations, although some time before the signing of the contract, all interested parties intensified their efforts to initiate the planning approval procedures. Various discussions took place between DFTG, Wilhelmshaven, and the licensing authorities (DGWE and WSB), but in late 1976 these were disturbed by an event that surprised all parties in the decision process, with the possible exception of the ME&T: the British company Imperial Chemical Industries Ltd (ICI) declared its interest in Wilhelmshaven as a site for a new petrochemical plant. Lower Saxony supported the ICI proposal and, since the only suitable site was one adjacent to that reserved for the LNG terminal, DFTG was informed that changes would have to be made to the jetty designs. The revised plans eventually agreed upon by DFTG, ICI, and Lower Saxony provided for two separate jetties, one for each facility, connected by a joint transport bridge, as illustrated in Figure 3.3. ⑤

Rounds C and D. Because ICI wanted to start construction of the petrochemical plant as soon as possible, this had a considerable impact on the dynamics of the decision process, and thus initiated round C in spring 1977. The new jetty design had hardly been settled when the Wilhelmshaven authorities initiated construction plan (*Bebauungsplan*) procedures for both projects, as required by federal construction law (*Bundesbaugesetz*). ⑥ Objections to the LNG terminal were then raised by a rather small group of environmentalists in Wilhelmshaven, the Hooksiel Citizens' Group, as well as by the municipality of Wangerland and some other local authorities on the grounds of safety risk, and the potentially negative effects on the environment, fishing, and tourism. Having heard and responded to these objections the Wilhelmshaven City Council approved the construction plan as proposed in July 1978. ⑩

A crucial first step in the realization of the LNG project was in June 1977, when the Dutch–German consortium (Ruhrgas AG, Salzgitter Ferngas GmbH, and NV Nederlandse Gasunie) signed a contract with the Algerian company Sonatrach for the sale of 8 billion m^3 of LNG per year for a period of 20 years starting in 1984, half of which was reserved for the German companies. (7) According to this contract, Sonatrach had to be notified of the exact location of the LNG terminal site by October 1978, a deadline which was later to cause problems.

In September 1977 DFTG filed applications for two licenses: the district government of Weser-Ems (DGWE) was responsible for approval of the land-based facilities through a preliminary license (*Genehmigungsvorbescheid*) according to the Federal Immission Control Law,[5] (8) but the harbor facilities were subject to another licensing procedure, the so-called plan specification (*Planfeststellung*), in accordance with the Federal Waterway Law. The Water and Shipping Board North-West (WSB) was responsible for granting this license, and considered the two jetties and the transport bridge (to be built by Lower Saxony, represented by the DGWE) jointly in three similar procedures.[6] (9)

Both DGWE and WSB then scrutinized the plans, and the application was laid open to the public. However, toward the middle of 1978, the WSB revealed a growing reluctance to approve the project because of their concern over unresolved problems related to shipping hazards in Jade Bay. The WSB therefore informed the Federal Minister of Transportation (FMT) that it did not wish to take sole responsibility for decisions regarding appropriate safety measures and the acceptability of the public safety risks, and so asked the minister to consider the critical questions himself. (11) The FMT was advised by a working group at the WSB (which included representatives of all relevant local authorities and regulatory agencies) and a committee of experts from the ministry. After consultations with other federal ministries, the FMT decided that the public safety risk would be acceptable only if a number of special measures were taken, the most important being an expensive change to the configuration of the deep-water shipping channel to the terminal. (12) This paved the way for the WSB to approve the plans for all the harbor facilities (DFTG and ICI jetties, transport bridge) between March and July 1979. (13)

When, shortly afterwards, the district government of Weser-Ems granted a preliminary license for the land-based facilities, (14) all potential obstacles had been surmounted. This was equivalent to an approval in principle of the whole project, allowing construction to start. (15) As for the current state of the project (end of 1982), construction of the terminal has not begun because of an unanticipated change in Algeria's export policy concerning LNG in early 1980. It is expected that construction will be delayed until the gas companies are able to negotiate a new LNG import contract.

PARTY CONCERNS

Table 3.2 gives an overview of the features and potential impacts of the proposed LNG terminal that were of concern to the main interested parties.

One has to be cautious in interpreting this table since the reasons for the parties' concerns about specific aspects of the project differed quite markedly. For example, the vital interests of some parties were directly affected, some parties had a legal mandate to look at particular consequences, while others used their concerns as strategic arguments to support their positions. While it is easy to identify explicit responsibilities, it is much more difficult to distinguish between "real" and "strategic" concerns. In addition, this list does not aim to assess in detail the roles of all the main parties, but rather to give a general picture of the variety of dimensions considered.

National and regional aspects were not considered in any great detail; most of the parties focused on local impacts. Consequently, the public debate dealt mainly with expected economic benefits or costs, potential risks, or negative environmental effects. In general, the parties most directly affected at local level had the broadest concerns, whereas the perspectives of the licensing authorities were generally confined to a few dimensions.[7] Of course, the picture is incomplete unless one takes the time dimension into account, and this will be done below.

THE MAMP FRAMEWORK

To illuminate some of the "substance" and the dynamics of the decision process, we introduce here the multi-attribute, multi-party (MAMP) framework (see Chapter 2). For the purpose of adding some structure, we have divided the process into a series of four rounds, which were discussed above. Each round is initiated by certain decisions or events that lead to an interaction phase. A round is concluded by a set of other events or decisions. Party interactions are characterized by the main concerns and the associated arguments of each actor. In so far as different questions are treated simultaneously, rounds also overlap. Full details of the rounds are given in Table 3.3.

Policy questions

The LNG siting decision in the FRG can be divided into three policy questions:

- Is a domestic LNG terminal desirable?
- If so, where should it be located?
- Under what conditions can the terminal be approved at that site?

It is apparent that the first two questions played a negligible role in the political decision process: the question of desirability was addressed only in the narrow sense of whether the LNG terminal was suitable for Wilhelmshaven. The siting question was not dealt with by public authorities except for the choice between different plots at Wilhelmshaven. A more general evaluation of the question of desirability and appropriate siting was only done within Ruhrgas and Gelsenberg (DFTG). Government

Table 3.2. Party/concern matrix.

Concerns		Parties								
		Applicant	Government							Other
		DFTG, Ruhrgas, Gelsenberg	Federal		State (*Land*)		Local			
			FMT	WSB	ME&T	DGWE	Wilhelmshaven	Wangerland	Friesland	HCG
National										
Need	N_1	x			x					
Import policy (diversification of sources)	N_2	x			x					
Regional										
Industrial development	R_2				x	x		x	x	x
Economic costs	R_3				x	x				x
Support of smaller gas supply companies	R_4				x					
Local										
Amenity and environment:										
land use, noise, air pollution	L_{11}	x		x		x	x	x	x	x
fishing, marine	L_{12}	x		x		x	x	x	x	x
tourism, recreation	L_{13}					x	x	x	x	x
Economic benefits:										
tax revenues	L_{21}				x	x	x			
jobs, short-term	L_{22}				x	x	x			
jobs, long-term	L_{23}				x	x	x	x	x	x
jobs, indirect	L_{24}				x	x	x	x	x	x
spin-off industry	L_{25}				x	x	x			
harbor activity and prestige	L_{26}				x	x	x		x	

Economic costs:										
facility	L_{31}	x								
harbor	L_{32}	x			x	x				
channel	L_{33}	x	x	x	x					
negative effects on fishing and tourism	L_{34}							x	x	x
Risks to population:										
shipping	L_{41}	x	x	x			x	x	x	x
gas storage	L_{42}	x				x	x	x	x	x
Risks to workers	L_5	x		x		x				
Ease and safety of shipping	L_7		x	x						
Applicant										
Profit	A_1	x								
Control over sources	A_2	x								
Image	A_3	x								

DFTG: LNG Terminal Company (Deutsche Flüssigerdgas Terminal GmbH)
FMT: Federal Minister of Transportation
WSB: Water and Shipping Board North-West
ME&T: Lower Saxony Ministry of Economic Affairs and Transportation
DGWE: District Government of Weser-Ems
HCG: Hooksiel Citizens' Group

Table 3.3. The MAMP framework applied to the decision process on LNG in the Federal Republic of Germany.

(*a*) ROUND A: 1972–July 1976.

I PROBLEM FORMATION

Assumptions: (1) Natural gas is an important source of energy and its benefits are generally accepted.
(2) The possibility exists to import Algerian LNG.
(3) The site at Wilhelmshaven is an area created to encourage industrial development.

Question: Given its feasibilty, is the proposed LNG project suitable and desirable for Wilhelmshaven?

II INITIATION

Ruhrgas and Gelsenberg (later DFTG) announce to Lower Saxony their intention to build an LNG terminal (1972). Wilhelmshaven is considered to be the most appropriate harbor by DFTG and ME&T. ① [a]

III INTERACTION

Party	Position	Arguments
DFTG, gas companies	For site at Wilhelmshaven	Need for natural gas (N_1) fits regional development plans (R_2); *technology is safe* (L_4).[b]
Wilhelmshaven	For site at Wilhelmshaven, subject to conditions on environmental impact	Contributes to industrial development (L_2); *safety and a high degree of environmental protection have to be ensured* (L_1, L_4).
ME&T	For site at Wilhelmshaven, subject to conditions on business structure	Beneficial to regional economy (R_2, L_2); support of gas supply companies in Lower Saxony (R_4).

IV CONCLUSIONS

1976: Lower Saxony and Wilhelmshaven commit themselves to support the project at the selected site; ② ③ the gas companies and DFTG agree on certain conditions (settlement contract). ④

[a] Circled numbers refer to the main events as illustrated in the PERT diagram (Figure 3.4).
[b] Symbols in parentheses are the same as those used in Table 3.1 to denote different concerns; arguments referring to population safety are indicated in *italics*.

Table 3.3. (*continued*) The MAMP framework applied to the decision process on LNG in the Federal Republic of Germany.

(*b*) ROUND B: 1972–spring 1977.

I PROBLEM FORMULATION

Assumption: For economic, technical and political reasons, Wilhelmshaven is a desirable site for constructing an LNG terminal. ②

Questions:
(1) Is the proposed LNG terminal acceptable with regard to safety and environmental protection requirements, town and country planning regulations, etc.?
(2) Which site at Wilhelmshaven is most appropriate?

II INITIATION

Lower Saxony and industry ask the regulatory agencies to give their views (1972/73).

III INTERACTION

Because of the type of interaction that went on at this stage of the decision process (informal consultations between companies and licensing authorities), and the restricted sources of information available, it was not possible to identify the individual positions and arguments of particular parties in this round. Therefore, we only list the parties involved and dimensions considered.

Parties

DFTG, gas companies, Wilhelmshaven, Friesland County, ME&T, DGWE, WSB.

Concerns

Safety and ease of shipping in the channel, nautical safety at the jetty.
Hydrological conditions and effects on the morphology of Jade Bay.

Population safety
Negative environmental effects: noise, air, and water pollution.
Technical and operational aspects.
Costs: facility, harbor, and maintenance of shipping channel.

IV CONCLUSIONS

(1) June 1974: Based on a siting analysis performed by a consultancy firm, DFTG selects one of the two plots offered by Lower Saxony. ③
(2) Spring 1977: DFTG, ICI, and Lower Saxony agree on plans to construct joint harbour facilities for the ICI petrochemical plant and the LNG terminal. ⑤

Table 3.3. (*continued*) The MAMP framework applied to the decision process on LNG in the Federal Republic of Germany.

(*c*) ROUND C: Spring 1977—July 1979.

I PROBLEM FORMULATION

Assumptions: Contract with Sonatrach ensures LNG delivery (deadline for the terminal site selection October 1978). ⑦ DFTG, gas companies, Lower Saxony, Wilhelmshaven are precommitted to a site at Wilhelmshaven. ICI want to build a petrochemical plant adjacent to the proposed LNG terminal, so Lower Saxony, ICI, and DFTG agree to combined jetty facilities. ⑤

Questions: (1) Is the project suited to local and regional development plans?
(2) Do the facilities meet all relevant safety and environmental regulations?
(3) Are there any private rights preventing approval of the project?

Process constraints: Various established statutory procedures for siting and licensing industrial facilities

II INITIATION

May 1977: Wilhelmshaven initiates construction plan approval procedures. ⑥ September 1977: DFTG requests approval of the DGWE to construct the facilities. ⑧ September 1977 to February 1978: DFTG, followed by ICI and Lower Saxony, request WSB approval to build jetty facilities. ⑨

III INTERACTION

Party	Position	Arguments
DFTG	For the project	Need for natural gas (N_1); supplies guaranteed (N_2); beneficial to region (R_2); improves infrastructure (L_{25}), jobs (L_{22}, L_{23}); *no population risks with safety measures* (L_{41}, L_{42}; 1—6)[a]; environmental regulations are met (L_{11}, L_{12}).
Hooksiel Citizens' Group	Against the project	Benefits not obvious; few jobs created (R_2, R_3); tourism and fishing affected (L_{34}); *population risks unacceptable due to probable vessel accidents and proximity of ICI and Mobil* (L_{11}, L_{42}; 8): psychological threat, pollution unacceptable (L_{11}, L_{12}).
Wilhelmshaven	For the project, with additional safety measures	Beneficial to industrial development (L_2, L_{25}, L_{26}); tax benefits (L_{21}); few jobs created, but important (L_{22}, L_{23}, L_{24}); *technology is safe* (L_4); minor environmental effects with safety measures (L_{11}, L_{12}); benefits outweight economic and environmental costs (L_1, L_2, L_{34}).
Wangerland and Friesland County	Against the project	Negative effects on environment (L_{11}), tourism, and fishing (L_{34}); *high risks due to vessel accidents and proximity of ICI and Mobil* (L_{11}, L_{42}).
DGWE	For the project	Beneficial to region (R_2, L_2); jobs (with ICI) (L_{22}, L_{23}, L_{24}); tax benefits (L_{21}); *no danger to public since worst conceivable accident not serious* (L_{42}; 11, 13); environmental regulations are met (L_1); negative effects acceptable (L_1, L_2, L_3).
WSB	For additional safety measures, but still undecided	No effects on Jade Bay (L_{12}); *non-neglible population risk* (L_{41}; 1, 3, 6, 9); *shipping safety not guaranteed* (L_6).

IV CONCLUSIONS

(1) July 1978: City of Wilhelmshaven approves construction plan. ⑩

(2) July 1979: DGWE grants license, subject to a number of technical requirements. ⑭

(3) WSB remains undecided with respect to shipping safety. ⑪

[a] Numbers in parentheses refer to the respective numbers of expert studies (see Table 3.4) used to support the arguments.

Table 3.3. (*continued*) The MAMP framework applied to the decision process on LNG in the Federal Republic of Germany.

(*d*) ROUND D: **August 1978–July 1979.**

I PROBLEM FORMULATION

Assumptions: (1) WSB is concerned about the safety aspects of LNG shipping in Jade Bay.
(2) Industry and Lower Saxony are concerned about delays in the decision procedure.
(3) Extensive risk-reducing measures (change in configuration of the deep-water channel) are proposed.

Questions: (1) Are the risks of LNG shipping in Jade Bay acceptably low in view of the proposed safety measures?
(2) Are the risk-reducing measures justified?

II INITIATION

August 1978: Federal Minister of Transportation (FMT) intervenes in the WSB approval procedure. ⑪

III INTERACTION

Party	Position	Arguments
Advisory Committee at FMT	Approve project, subject to safety conditions	*Limited population risk is acceptable* (L_{41}; 17).

IV CONCLUSIONS

(1) March 1979: The FMT decides that the residual risk of the LNG terminal is acceptable.[a] ⑫
(2) July 1979: The WSB approves the license for the LNG terminal. ⑬

[a] Residual risk is defined as the risk remaining after all safety measures have been enforced.

agencies influenced the answers to these questions only in a very indirect way through energy (Federal Minister of Economic Affairs) and regional development policies (ME&T). One could therefore say that the most important part of the decision had been taken before the first round considered here began. This study does not account for these crucial predecisions because either they were not made in the context of LNG, or were taken internally to an organization, and thus outside the political arena.

The first two rounds can be interpreted as preparatory stages to the central part of the public decision-making process concerning the LNG terminal, i.e., the formal licensing and approval procedures that took place in round C. In round A, the questions addressed were whether and to what extent Lower Saxony and the local authorities were willing to support the project in general, and were willing to carry out planning and construction approval procedures. Because of this restricted agenda, the party interactions had the character of negotiations and consultations based on common interests, such as industrial and economic development. The local residents knew of course about these activities, since DFTG's proposal to construct an LNG terminal at Wilhelmshaven had been reported in the media; but details of the project were treated as confidential during the first two rounds. By the end of round A, the DFTG and the public authorities involved were committed to a particular site and to certain preconditions for the realization of the project.

In round B the regulatory agencies, particularly the licensing authorities WSB and DGWE, had to give their preliminary views of the feasibility of the project, and they supported the parties of round A in their choice of possible sites at Wilhelmshaven. Although this interaction was informal and not binding, mutual information and internal deliberations had an important influence on both DFTG's preparation of the plans and the way in which the license applications were later evaluated. The DFTG was informed of the main problems from the perspective of the regulatory agencies, and was thus able to improve in advance the weak points in the plans. The licensing authorities were also able to collect and process technical information and to set the conditions that the DFTG would have to fulfill in order to comply with the relevant regulations.

Site approval formed the agenda of round C, since the site had already been selected. Various aspects, such as regional development, safety, and environmental impacts were considered in round C. While the issues at stake were quite similar to those of round B, the main difference was that the formal procedures of round C were legally binding, thus enabling new parties to become involved in the decision process. For example, only after detailed design plans were published could the public air its views on the proposed LNG terminal. In other words, only *after* important predecisions such as the choice of an appropriate site and the elaboration of a desirable terminal design had been taken (during rounds A and B) did those directly affected by the project have an opportunity to participate in the debate.

Interaction between the interested parties therefore became considerably more adversarial during round C, although not in the form of open public debate because of legal constraints. Concerns about the project could only be raised within a certain period of time and within the problem bounds defined by law for each of the procedures. Town and country planning, for instance, does not enable safety or environmental issues to be addressed in any detail, even if the procedure is adjusted to a particular project, as it was in this case.

Local opposition to the LNG terminal was aroused during the spring and summer of 1976 when the public and neighboring municipalities were officially informed about the project as part of the legally required construction plan approval procedures. It had been known for some time that the area to the north of Wilhelmshaven had been earmarked for industrial development; but the reaction to the actual DFTG/ICI plans could best be described as "planning shock". The residents of Hooksiel were under the impression that the area had been zoned for small-scale, environmentally harmless industries since it had been designated as a "restricted industrial zone", but this injunction in fact referred only to noise levels. The hazards and potential environmental effects became more obvious after the publication of the plans, and offered a clear focus for the opposition.

The opposition to the DFTG/ICI project was made up of a coalition of environmentalists in Wilhelmshaven, inhabitants of Hooksiel who felt directly affected, and local authorities (Wangerland and Friesland) who considered the plans a serious threat to the economic development of Hooksiel. This opposition did not include groups with general anti-industrial, anti-capitalist ideologies, nor "grassroots" movements that tend to refuse to cooperate in principle with established political institutions. The local residents' concerns about risk, negative environmental effects, and damage to local businesses and tourism were closely related — indeed, sometimes interwoven.

Despite its efforts to support these concerns with as many arguments as possible, the opposition was not very successful in fighting its case. During the construction plan approval procedure the Wilhelmshaven City Council overruled most of the objections on procedural grounds, referring to the following licensing procedures. The licensing authorities considered themselves the wrong audience for issues they felt were not technical but political in nature (such as insufficient public participation, no long-term economic benefits, etc.). Only when Wangerland and the Hooksiel Citizens' Group (HCG) each appealed to the courts against the ICI plan were the Wilhelmshaven authorities and Lower Saxony willing to consider a compromise. HCG's case was dismissed, and Wangerland voluntarily withdrew after Lower Saxony committed itself to higher subsidies towards tourist facilities in Hooksiel as a sort of compensation.[8]

Numerous discussions took place between the DFTG and the licensing authorities, and the public authorities examined numerous technical details. While advice from independent experts had already been used in the preparatory stages of the procedures, most of the detailed technical reports were only commissioned during round C. By the end of this round preconditions and requirements for the approval of the project had largely been settled.

The one remaining question — namely shipping risks in Jade Bay — set the bounds for round D, and thus limited the number of participating actors. For the most part the debate took place within federal government ministries, so that the final decision can be ascribed to a single decision maker, the Federal Minister of Transportation (FMT) during round D. However, he could not overrule decisions that had already been made in previous rounds, so his power was somewhat limited.

UNUSUAL FEATURES

A number of features and events in the site decision on the first LNG terminal in the FRG can be considered remarkable or unusual.

(1) The preparatory phase of the approval procedures stretched over more than five years from the first involvement of the public authorities to the lodging of the formal applications. This was longer than in other industrial siting decisions of comparable importance, such as the ICI project.

Because of the informal character of talks between industry, local authorities, and the state industrial development agency (ME&T), there was no prescribed limit to their duration; the same applied to the preparatory phase of the approval procedures. Several factors contributed to the length of this phase, such as Lower Saxony's requirement that smaller regional gas companies could join the corporation operating the LNG facility, which was a serious point of contention in the negotiations with Ruhrgas and Gelsenberg. The problem was resolved by a compromise reached in 1976 restricting the overall share of these companies (to be nominated by the ME&T) to 26%, and this was executed in 1979. Furthermore, the pricing negotiations between the gas companies and Sonatrach were very tough; in fact a pre-contract signed in 1974 was not effected because of differences on this point. Another important reason for the delay was the public authorities' lack of experience with LNG technology; the DFTG thus had to reply to the numerous doubts about its acceptability, particularly in relation to public safety and environmental impact.

(2) The planning process, public discussions, and formal procedures were considerably affected by the planned close proximity of the LNG terminal and the ICI petrochemical plant.

Some of the consequences of this proximity have been discussed above; one of these — apart from the necessary revision of the jetty design — was that Lower Saxony took a more active role since it became an applicant itself. Some citizens of Wilhelmshaven and neighboring municipalities were alarmed by the new plans, although they had already known about the LNG terminal. The licensing authorities, particularly the WSB, then had to cope with a much more complex technical situation, since hazards could result not only from incidents in each of the facilities or on the respective tankers, but also from "domino-effect" interactions between the two activities.[9]

(3) An important part of the approval process, particularly town and country planning procedures, had to take place within considerable time constraints.

Perhaps the greatest impact of the ICI project on the LNG terminal decision was the increased momentum that resulted from the time constraints imposed. ICI (fully supported by Lower Saxony) was pushing for an early decision on the petrochemical plant, so that a number of decisions, such as the construction plan approval for ICI, were taken precipitately, creating legal difficulties,[10] and provoking conflicts at local level. In mid-1978 DFTG also raised concerns about the delays since Sonatrach had stipulated that the terminal location must be announced by October of that year. In retrospect, this seems to have been mainly a tactical argument since it would probably have been possible to postpone this deadline without negative consequences.

(4) The WSB's doubts about making a decision on the acceptability of the risks imposed by LNG tanker traffic led to the unusual, though legally correct, involvement of a federal ministry in the approval process.

The pressure of time acknowledged by representatives of the WSB and the Wilhelmshaven authorities leads us to a key event in the decision process: the inability or unwillingness of the WSB to approve the project or, to be more precise, to decide independently on the acceptability of the risks of LNG shipping. A most interesting aspect of this event from our perspective is the extent to which the WSB's attitude can be explained by the type of technical information that the board had to consider, and the manner in which the data were perceived and interpreted.

We will therefore put the risk issue into perpective by accounting for other possible influencing factors.

- The WSB did not employ professional experts in the field of hazardous materials, and so had to rely to a large extent on outside experts.
- Difficulties resulting from the above point must have been aggravated by the time constraints.
- There are indications that the "human factor" (i.e., personal ambitions or biased problem perceptions of key individuals) played a part. When interviewed, representatives of some interested parties mentioned several times (in reference to the WSB) the influence one person could have on the behavior of an agency. If such an influence existed, it was in the direction of risk avoidance.
- The WSB felt the need to involve the FMT in a decision concerning the possibility of a catastrophic accident, in order to spread the responsibility for that decision. In addition, the WSB might need federal support for several unusually far-reaching and expensive measures it was going to require from the applicants.

(5) Local opposition to the LNG project was shaped to a large extent by the location of the terminal in that those who would have to bear the main environmental costs were not the inhabitants of Wilhelmshaven, but those of the neighboring municipality of Wangerland.

Because of the terminal's location, it was very difficult for the decision makers to take into account the various concerns debated at the local level. The approved plans for the terminal did incorporate a number of important revisions made in response to objections raised during the licensing process. For instance, the revised storage tank design (agreed after thorough deliberations between DFTG, DGWE, and Wilhelmshaven) removed an important point of criticism. It was clear, however, that Hooksiel and Wangerland would have been satisfied only if the plans to build the terminal had been dropped altogether; a compromise was eventually reached because these parties felt that local political and financial resources were not sufficient to defeat the project. Also, Wangerland was afraid of losing Lower Saxony's financial support for its tourist development plans if it did not accept this solution.

THE RISK ISSUE

Safety questions were considered at all stages in the decision process, but under changing assumptions and constraints. The institutional risk evaluation procedure followed essentially established paths, although the final stage of decision making was quite unusual. With respect to the technical risk assessments, the licensing authorities considered that their approach to evaluating the public safety risk from LNG technology exceeded standard approval procedures and established a precedent for comparable industrial siting decisions.

Ruhrgas and Gelsenburg (DFTG) familiarized themselves with the main safety problems of LNG technology before entering the site selection process (pre-round A). During their negotiations with the state and local authorities (round A), public safety risk was not the predominant issue, but it nevertheless played an important role because the acceptability of the risks was a necessary condition for the approval of the project. In particular, the Wilhelmshaven authorities needed confirmation on this point since its constituents would be directly affected by the hazards of the terminal.

Consequently, the question of risk was discussed at length by the DFTG and the public authorities concerned with the project. The regulatory agencies were asked to give their views on the acceptability of the risks related to LNG (round B); they deemed the project feasible in principle, but only if more detailed research revealed no unexpected difficulties. Their evaluation was based on information obtained from comparable authorities in other countries (including the Dutch TNO report; see Chapter 4), official technical boards, and company experts, but not on a comprehensive risk analysis done specifically for Wilhelmshaven.

Studies concerning public safety risk (*Gutachten*) carried out by certified experts[11] were only introduced in rounds C and D. These were commissioned by (i) the DFTG, to support its applications for the two licenses;

(ii) interveners (Mobil Oil, HCG), to support their objections to the project; and (iii) the licensing authorities (DGWE and WSB), to advise them and to justify their decisions against potential claims from those directly affected.

Most of these *Gutachten* did not address the problem of population risk directly, but considered more piecemeal issues such as technical safety, safety of shipping, or prevention of and mitigating measures against fire and explosions. The focus on safety instead of risk has to be seen against the background of the relevant legislation in the FRG, according to which public authorities are responsible for protecting citizens against "harmful environmental effects and other dangers, considerable detriments, and considerable disturbances" (Federal Immission Control Law, paragraph 5). Table 3.4 gives an overview of the expert studies dealing with population risk.

Because of their importance in the final decision on the LNG terminal, we now briefly describe the studies made by Brötz and Krappinger. In its deliberations with the applicant, the WSB came to the conclusion that ship collisions and groundings of LNG tankers could have disastrous consequences, so that estimates of the expected frequency of such events were deemed necessary. This work was performed by Professor Krappinger, head of a shipbuilding research institute (Hamburgische Schiffsbauversuchsanstalt GmbH), who combined historical records of vessel accidents in Jade Bay and a computer model for shipping operations with estimated rates for LNG tank rupture in the event of an accident. Using these data, Krappinger calculated probabilities on the order of 10^{-3} for a major spill per year (Krappinger 1978a, b). These probability estimates provoked much criticism from licensing authorities and industry experts, who thought that the assumed spill rates were far too high. In the final version of his report, at the request of the WSB, Krappinger modified his assumptions so that the probability of a tank rupture was lower.

Professor Brötz, an expert frequently consulted by licensing authorities, was asked by DGWE and WSB to scrutinize the technical safety (*Sicherheitstechnik*) aspects of the applications for both licenses. In particular, the WSB requested investigation of several types of potential major accidents (Brötz 1978). Brötz was also asked to assess the likelihood and consequences of a maximum credible accident at the land-based facilities, and to give his opinion on the safety of the revised design for storage tanks, which now incorporated an outer tank of reinforced concrete instead of steel (Brötz 1979). An important part of his two reports dealt with numerous technical devices, design requirements, and operational prescriptions to prevent failures and confine accidents. After this traditional engineering approach, Brötz also considered (at the request of the WSB) the potential physical consequences of accidents to the public and the neighborhood. While potential hazards to the population, particularly at the recreational zone near Hooksiel, could not be completely excluded

Table 3.4. List of expert studies dealing with public safety in the LNG decision process in the FRG.

No.	Author	Date of completion[a]	Commissioned by	Type of author[b]	Topic	Results	Methods used to assess risk
1	ICT -Fraunhofer Association[c]	1977, Dec.	DFTG, applicant	Publicly subsidized institute for applied research	Ignition and dispersion properties of natural gas/air mixtures	Safety for neighborhood ensured	No information
2	Engler-Bunte Institute I[c]	1978, Jan.	DFTG	University institute	Safety questions regarding gas and fire control technology of land-based facilities	Safety ensured; prevention measures appropriate	No information
3	Engler-Bunte Institute II[c]	1978, April	DFTG	University institute	Safety questions regarding gas and fire control technology for harbor facilities	Safety ensured; prevention measures appropriate	No information
4	Böttcher/ Rother I[c]	1978, Feb.	DFTG	Certified experts	Scrutiny of terminal plans (land-based facilities) with respect to proposed fire safeguard measures	Measures sufficient	No information
5	Böttcher/ Rother II[c]	1978, April	DFTG	Certified experts	As above	Measures sufficient	No information
6	Germanischer-Lloyd (1978)	1978, March	DFTG	Semi-official/ insurance company	Safety of LNG tankers	High safety standard ensured; LNG spills on Jade Bay not expected	Deliberation of potential accidents (qualitative)
7	Energy Analysts (1978)	1978, June	Mobil Oil AG	Consulting firm	Scrutiny of application regarding the quantification of potential hazards	Risk analysis should be completed	No independent analysis performed
8	Johannsohn (undated)	Undated	Hooksiel Citizens' Group (HCG)	Certified expert	Hazards of LNG and liquid chemical spills related to tanker accidents	Selected survey of the literature	No independent analysis performed
9	Krappinger (1978a, b)	1978, June	WSB	University professor/ research institute	Estimated frequency of tanker accidents in Jade Bay and at the jetties	Probability of tanker accidents leading to an LNG spill on the order of 10^{-3} per year	Probabilistic calculations of critical events based on historical data and shipping simulations
10	Karlsch/Spohn[c]	1978, July	Wilhelmshaven (as agency of DWGE)	Certified experts	Prevention measures for local population related to fire hazards and catastrophes at harbor facilities	Measures essentally sufficient	No information

11	Karlsch[c]	1978, Nov.	Wilhelmshaven (as agency of DGWE)	Certified expert	Prevention measures for local population related to fire hazards and catatrophies at land-based facilities	Measures essentially sufficient	No information
12	Brötz I (1978)	1978, Dec.	WSB	University professor/ certified expert	Scrutiny of application regarding safety technology; analysis of selected potential accidents: harbor facilities	No dangers in the sense of of the relevant laws, if additional safety requirements are regarded	Quantitative/deterministic with respect to physical consequences of accident; qualitative estimation of probability of failure events
13	Brötz II (1979)	1979, July	DGWE	University professor/ certified expert	Scrutiny of application regarding safety technology; analysis of selected potential accidents: land-based facilities	No dangers in the sense of relevant laws, if additional safety requirements are regarded; new storage tank design desirable	As above
14	TÜV (1979)	1979, March	WSB	Body constituted under public law	Scrutiny of application with respect to detailed technical safety regulations (mainly concerning occupational safety)	Essentially consistent with relevant regulations	Qualitative discussion of technical details
15	BAM (1979)	1979, March	WSB	Federal technical institute	Review of plans and requirements with respect to appropriateness, state of the art, comment on WSB report	Safety measures sufficient, residual risk acceptable	Qualitiative discussion of probability and consequences of major accidents
16	WSB (1978)	1978, Oct.	FMT	Licensing agency	Risk assessment and discussion of safety measures relating to tanker traffic with hazardous chemicals to the DFTG/ICI jetties	Additional safety measures have to be taken; even then the residual risk to the population remains considerable	Qualitative evaluation of other expert studies
17	Advisory committee at FMT (Risikoabschätzung 1979)	1979, Jan.	FMT	Permanent advisory committee at FMT	Risk assessments relating to shipping and transfer of LNG and hazardous chemicals on Jade Bay	Given the proposed safety measures are taken the residual risk is acceptable	Qualitative evaluation of other expert studies including WSB report

[a] First version.
[b] All authors apart from state agencies were either certified experts or they had a comparable semi-official status.
[c] These reports were not available to the author; information on them was mainly drawn from: BAM (1979), DFTG (1978), DGWE (1979), and WSB (1978, 1979).

from these calculations, Brötz considered them to be minor enough in view of their very low probabilities to comply with legal requirements.

In a standard approval process for a large industrial facility, expert studies such as these, together with those dealing with purely environmental issues, would have been sufficient to prepare and support the final decisions of the licensing authorities. In fact, the preliminary license application, according to the Federal Immission Control Law, was terminated at this stage (DGWE 1979). However, the situation became more complicated and difficult with the plan specification procedure, so that another set of risk assessments had to be performed before final approval could be given (round D).

By summer 1978 someone at the WSB apparently became concerned about the LNG shipping risk. From Krappinger's study it was clear that accidents involving tankers transporting LNG or hazardous chemicals were the greatest potential risk, and had a significant probability of occurrence along the shipping channel or at the jetty. Since the consequences of such accidents "with respect to their scope and likelihood" had not been assessed in a way that satisfied the WSB, the board was therefore not prepared to approve the terminal without support from federal level.

One of the subsequent measures of the FMT was therefore to establish a working group at the WSB to "intensify" the procedure and to elaborate a decision aid for the board.[12] Based on discussions within this group, the WSB drew up a report for the minister which, owing to time pressure, reflected mainly its own views (WSB 1978). The report explained why the information on the question of population risk was not considered to be comprehensive and reassuring enough to warrant approval of the terminal. The board also felt that the expert studies disagreed substantially even on fundamental questions. Furthermore, there were differences between the experts of the WSB working group regarding methods of evaluating risks. Nevertheless, after proposing and deliberating various risk-reducing measures the WSB tried to evaluate the residual risk of LNG tanker traffic by reviewing and comparing the different expert studies. Despite the fact that the probability of a major accident was deemed to be very low, the WSB came to the conclusion that the population risk was significant because of the serious consequences of such an accident.

The FMT handed this report, together with all expert studies, over to the Advisory Committee for the Transportation of Hazardous Goods, a permanent board of experts at the ministry. The committee formed a working group to prepare a final risk assessment. Four of the five members of this group belonged to institutions that had already been involved in the decision process.[13] Whether it was intended to or not, the composition of this group, according to one ministerial civil servant, played an essential role in establishing some consensus between experts who thus far had not agreed on such important points as an appropriate method of assessing risk. After discussing methodology and qualitative factors associated with several major hazards, the working group concluded that there was a residual risk related to shipping LNG and hazardous chemicals; this was smaller in terms of probability but larger in terms of potential consequences than comparable existing risks. This residual risk would be acceptable "if public authorities took the responsibility for it, taking into account political and economic benefits of the proposed project" (Risikoabschätzung 1979, p12). This view was adopted by the FMT and paved the way for the plan specification approval, including its injunctions and requirements for the applicant (WSB 1979).

Technical analyses and expert studies dealing with risk to life and limb were thus widely used in the decision process. All of these were introduced in the context of formal approval procedures, but the site selection decision was made without a comprehensive risk analysis that was available to more than one of the interested parties. These studies were related to their use in the licensing procedures; the problems discussed were well defined but their scope was often narrow, and authors of the studies were selected on the basis of their professional qualifications, reputation, and official or semi-official status. In most cases the presentation of the results was oriented towards the general mandate of the regulatory agencies to prevent dangers to the public and neighborhood, or, more specifically, to ensure compliance with particular safety regulations. The question of risk acceptability was phrased and answered within this framework only.

In general, the risk studies were used to justify specific actions of their commissioners, rather than to advise, as is illustrated by the timing of the studies. For example, the Brötz report was completed before the two main licenses were granted, but *after* the objections raised during the proceedings had been dealt with in public hearings. It would therefore have been surprising if the findings of the expert studies had influenced the decision-making process significantly.[14] However, for particular aspects, such as the probability of shipping accidents with serious consequences, the reports might have had a significant influence on the main parties' views and positions. At least the results of the Krappinger *Gutachten*, unlike other reports, were published before the public hearings had taken place, and were used by the WSB to support their main arguments. Also most proposals regarding improvements, additional safety measures, and construction and design features discussed in the expert studies, were taken into account in the final plans. The most prominent example of this was the modified design of the storage tanks.

In trying to evaluate the way in which expert opinion was introduced into the decision process, the task of assessing the population risk was accomplished without high costs in terms of money or time. Problems arising from contradictory expert views that had caused some trouble during the licensing process at the WSB were effectively resolved by the FMT, in particular by forcing dissenting experts to work together to determine the acceptability of the population risk. However, this final risk assessment did not include a thorough review of the studies made by other experts in the field.

CONCLUDING REMARKS

From a broader perspective, one of the shortcomings of the decision process investigated in this chapter was the secrecy and inaccessibility of information in the early stages of decision making, before the formal procedures began. In an atmosphere of growing societal concern about risk and the negative effects of technological development it is advisable to reconsider the concept of the strict isolation of preliminary, "in principle" decision making in a political setting without public debate on safety questions, and the existing licensing procedures where safety issues can only be debated in the context of a preselected site. Otherwise planning shock and subsequent strong reactions by the public, as in this case, are only to be expected.

Much of the concern at the local level was related to procedural aspects, although the opposition was not very strong in Wilhelmshaven. Some problems could probably have been avoided by a broader debate on safety questions at a stage when the decision was still open (i.e., during the zoning procedure). However, decision makers in the FRG, as well as in other countries, often disagree with this view. For instance, Anthony Barrell of the Health and Safety Executive in the UK has commented:

> Usually when you give more information you in fact arouse more controversy, not less. But nevertheless, I think it is something we should aim to do. (KLS 1982, p488)

The type of conflict about the LNG terminal and the composition of the local opposition explains to some extent why the issue provoked so little public protest. For instance, there was no participation by national or regional environmentalist groups; among other factors, this can be attributed to the concentration of the existing opposition on purely local issues, risk being only one of several areas of concern.

In societal debates comparable to that on nuclear power (which in the future could also embrace issues like LNG terminal siting), more comprehensive risk assessments and public participation earlier in the decision-making process might be recommended, not only to increase the acceptance of technological developments, but also to reduce the financial risk of a possible rejection or major design change of a project at an advanced stage of planning. On the other hand, the public authorities' current practice of requesting only narrowly defined information from a number of experts, and then integrating these fragmented views into a more balanced and complete assessment, can be seen to have important advantages too. Compared with having comprehensive risk assessments performed by outside experts, this practice helps to ensure a higher degree of responsibility on the part of the public authorities and thus makes the decision process, at least in principle, open to a certain degree of political control. The limitations of this way of handling public saftey risk have to be seen in the ever-growing complexity of technological issues that might at some point go beyond the expertise of public employees, such as those in charge of the licensing procedures related to LNG facilities.

NOTES

1. For a discussion of the safety risks related to this technology, in particular distant vapor cloud ignition following an LNG spill, see Chapter 7.
2. Other parties that participated at some stage in the decision process were:
   - Imperial Chemical Industries Ltd, who had applied for planning permission to build a petrochemical plant in the vicinity of the LNG terminal.
   - The Federal Minister of Economic Affairs, who evaluated the economic benefits of the project after it had become an issue at federal level.
   - The Military District Administration II in Hannover, concerned about possible effects of LNG tanker traffic on the NATO naval forces based at Wilhelmshaven.
   - Wangerooge and Schortens, two other neighboring municipalities.
   - Several citizens' action groups and individuals in Wilhelmshaven concerned about environmental issues.
   - The Mobil Oil Company, which operates a refinery to the south of the planned terminal, raised concerns about shipping hazards and potential cost of keeping the jetty area clear of sand deposits.
   - Local regulatory agencies, such as the Wilhelmshaven Harbor Authority and the Factory Inspection Agency in Oldenburg.
   - The Brotherhood of Harbor Pilots Weser II/Jade, a kind of trade union.
   - Several official technical boards.
   - Various consultants.
   - A number of interest groups such as aquatic sports clubs and a local fishermen's trade union.
3. Niedersächsischer Minister für Wirtschaft und Verkehr, at the time called Minister für Wirtschaft und öffentliche Arbeiten.
4. Commitments from both sides on issues such as the provision of physical infrastructure, subsidies to be granted, support in approval procedures (public authorities), the types of investment to be made,

environmental protection measures, and the possibility of smaller gas companies taking over a share in DFTG (industry) were eventually laid down in the "settlement contract". Such a contract (*Ansiedlungsvertrag*) is legally binding under private law, and is frequently used in connection with industrial projects subsidized by public funds.

5. *Bundes-Immissionschutzgesetz.* The word "immission", in contrast with "emission", focuses on the potential detriment to the environment rather than indicating the source of pollution.

6. In addition, a procedure concerning water management and protection was carried out by the DGWE. Although it became clear that this license would be granted, the procedure was never terminated because of the delay of the whole project.

7. The DGWE seems to contradict this statement, but this is due to the fact that it was not only a licensing authority, but also assumed planning functions as a direct agent of the ME&T (see Table 3.1).

8. The cost of this recreational center, consisting of an enclosed seawater swimming pool with artificial waves and various communications facilities, was estimated at DM 12.5 million (approximately US$6 million), 80% of which would be covered by different state subsidies. This was significantly more than the usual rate of 50% state subsidies for economic development projects. Moreover, it was not clear that without this agreement this particular project would have been supported at all by the Lower Saxony authorities.

9. Such a domino effect would of course also be conceivable between the Mobil Oil refinery (to the south of the ICI site) and the LNG terminal. This point was raised and considered in one of the risk studies without, however, playing a significant role in the public debate (probably because the ICI petrochemical plant, technically and politically, was such a prominent issue).

10. This procedure had to be repeated twice owing to concerns raised about its legal correctness.

11. The technical term in the FRG is "gerichtlich beeideter Gutachter", meaning that the expert has been approved by the courts, and that he has agreed to judge impartially.

12. Apart from the WSB and the FMT, this working group consisted of the DGWE, Wilhelmshaven, Friesland, the Military District Administration II, and four technical boards, namely: the Federal Institute for Material Control (BAM: Bundesanstalt für Materialprüfung) and for Physics and Technology (PTB: Physikalische-Technische Bundesanstalt), the Federal Bureau for the Environment (Umweltbundesamt), and the Society for Nuclear Power Utilization in Vessel Con-

struction and Shipping (GKSS: Gesellschaft für Kernenergieverwertung in Schiffbau und Schiffahrt).

13. Germanischer-Lloyd, BAM, PTB, and the FMT.

14. A risk analysis performed for the ICI petrochemical plant had even more far-reaching consequences since it resulted in the expensive resettlement of almost 100 inhabitants of Inserhausersiel, a village adjacent to the ICI site and therefore particularly exposed to potential escapes of hazardous chemicals such as vinylchloride (Vahrenholt 1980).

Chapter 4

The Netherlands: The Rotterdam – Eemshaven Debate*

Plans to import liquefied natural gas into the Netherlands were first drawn up in the early 1970s and resulted in the initiation of studies and discussions on various aspects of LNG technology. The siting question, however, was not an urgent one until 1977, when a contract was signed with the Algerian company Sonatrach to import 4 billion m^3 of LNG per year, for a 20-year period starting in 1983. Following extensive political discussions at various levels, an LNG terminal site at Eemshaven, in the northern province of Groningen, was finally selected and approved by the Dutch cabinet and parliament in 1978 (Tweede Kamer 1978). This decision outcome was significant because Eemshaven only became a serious candidate in late 1977; detailed studies and policy advice to and within the government (including the cabinet) had previously focused on Maasvlakte in the Rotterdam harbor area, as the preferred terminal site (see Figure 4.1).

This review of the Dutch decision-making process involved in LNG terminal siting assesses the political factors that led to the final choice of Eemshaven. Although the Eemshaven site was approved in 1978, the Algerians have since canceled the LNG contract. As a result of the absence of alternative suppliers, further plans to construct an LNG import facility have been postponed.

SETTING THE STAGE

Decision making in the Netherlands is relatively centralized, with the national government coordinating major policy decisions concerning

*This chapter was written by Michiel Schwarz, based on a complete case study report (Schwarz 1982). The author worked closely with Joanne Linnerooth in integrating the case study material within the MAMP framework.

Figure 4.1. The Netherlands.

regional development, energy planning, and land use. Specific siting approval for industrial development, however, is usually a matter for local authorities, so that many planning decisions involve a combination of local and national procedures.

At the local and national levels, public participation in decision making takes place via elected councils of representatives. Dutch political tradition has emphasized the importance of pluralism in government and society, and this is reflected in the large number of political parties.

Parliamentary councils are elected by proportional representation, so that ample opportunity is provided for a wide range of political views to be aired. Against the background of the multi-party, consultative style of political decision making, the tradition of pluralism explains why such a large number of interested parties were involved in the LNG facility siting debate. The nature and complexity of the decision events reviewed here reflect this point.

As policy issues have become more complex and interrelated, there has been an increasing emphasis on interdepartmental coordination within the Dutch government. Where the responsibilities of different departments overlap, coordinating committees are often set up with the aim of producing agreement between top civil servants from different ministries, in order to prepare governmental (and usually cabinet) policies (Binnenlandse Zaken 1980). In the case of the LNG decision most of the coordination took place within the Interdepartmental Committee for North Sea Affairs (ICONA), under the responsibility of the Ministry of Transport and Public Works (ICONA 1978c).

The decision to import and store LNG at a domestic terminal was made, to a large extent, within the context of national energy policy by the semi-state-owned company NV Nederlandse Gasunie. Dutch official energy policy, first formulated in 1974, established the need to import natural gas — including LNG — in order to conserve Dutch gas fields and to maintain strategic domestic natural gas reserves (Tweede Kamer 1974). In the mid-1970s natural gas contributed over half of the total energy supply of the Netherlands (CBS 1978). This policy provided the mandate for major decisions to be made by Gasunie, as well as the policy framework for subsequent government activities in connection with LNG.

The *national* dimension of decisions concerning LNG is emphasized by the fact that Gasunie is not an independent entity; the state has a 50% stake in the company.[1] Gasunie was set up in 1963 and is responsible for all matters concerning the supply, management, sale, and distribution of natural gas (domestic and foreign) to Dutch consumers. Close contacts thus exist between Gasunie and the government, and these are formalized through the Ministry of Economic Affairs.[2]

The number of sites considered for the planned LNG facility was narrowed down at an early stage in the decision process. Until late 1977, intragovernmental evaluation of the technical feasibility and costs/benefits of various sites had led to strong support for Maasvlakte in the Rotterdam harbor area. Technical studies (such as those by the Netherlands Maritime Institute, NMI) had concluded that alternative sites were less feasible; in particular, Eemshaven harbor was declared unsuitable on nautical and technical grounds. The focus on Rotterdam was further reinforced by Gasunie, which had favored a Maasvlakte LNG site from the start for a number of economic and corporate-strategic reasons. The consensus between Gasunie and the Ministry of Economic Affairs on the preference for a Rotterdam site was an important element in defining the structure of the early rounds of the debate.

Located in the south of the country, Rotterdam is the closest major port to Algeria, the anticipated source of LNG, and is in close proximity to the major gas consumers (see Figure 4.2). Rotterdam also provides the benefits of a well established, international deep-water port where the annual turnover of goods exceeds 270 million tonnes, two-thirds of which involves oil and oil products (Rotterdam 1978a). In the early 1970s, however, Rotterdam officials started to respond to a major shift in Western Europe from oil to other fuels such as natural gas, LPG, and coal. They saw the need to avert a decline in the harbor's economic activities by developing facilities to handle alternative energy products. Such a transition would take advantage of the existing infrastructure and labor surplus, and would help to maintain Rotterdam's position as Europe's major energy distribution center.

However, due to the lack of consensus among the three major local authorities with jurisdiction over the Rotterdam sites, and because of the threat of delay and imposed conditions, in December 1977 Gasunie reintroduced Eemshaven as a possible site for the LNG terminal (see Figure 4.3). New technical studies on the nautical conditions at Eemshaven concluded that recent shipping movements had made the approach to the harbor suitable for LNG-type tankers, and so Eemshaven was officially proposed. At the time, the government had already singled out the northeastern region as a focus for development and had initiated plans to attract industry. Dutch regional development policies specifically favor plans that provide a more equitable distribution of land use, economic activities, and employment, so that siting the LNG terminal in Eemshaven was viewed by the local authorities as a means by which the government could demonstrate its commitment to promoting industrial activities in this region. Based on the perceived socioeconomic advantages, the Groningen local authorities successfully organized a large number of public and private interests to lobby in support of the facility in Eemshaven.

In addition to the land-based sites at Rotterdam and Eemshaven, several other alternatives were also considered. These included an artificial island 27 km offshore, connected by pipeline to Maasvlakte or other parts of the mainland, and an offshore tunnel terminal system (OTTS) comprising a receiving platform 4 km offshore (11 km to the nearest town of Hoek van Holland), connected by an underwater pipeline to a storage site at Maasvlakte. Another "intermediate" solution was also rejected — the Voornedam breakwater, a 7—10 km long dam extending from the southwestern point of Maasvlakte. These three alternative solutions had the advantage that the shipping routes to the unloading terminals would not interefere with other Rotterdam harbor traffic, but they were rejected by

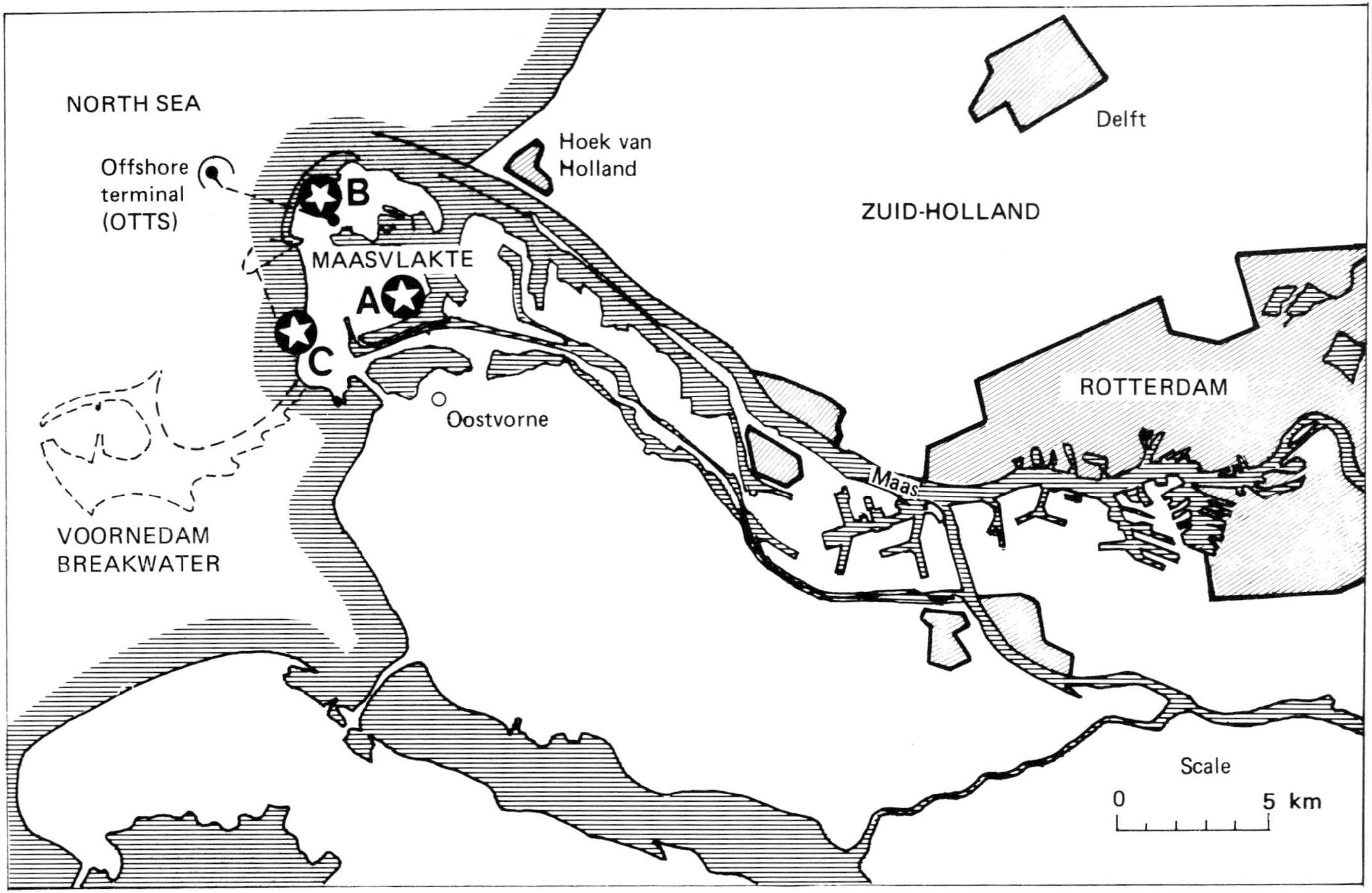

Figure 4.2. The Rotterdam harbor area, showing the three proposed Maasvlakte sites (A, B, C), Voornedam breakwater, and the offshore tunnel terminal system (OTTS). In addition, there were five different proposals for a North Sea island location for the LNG terminal.

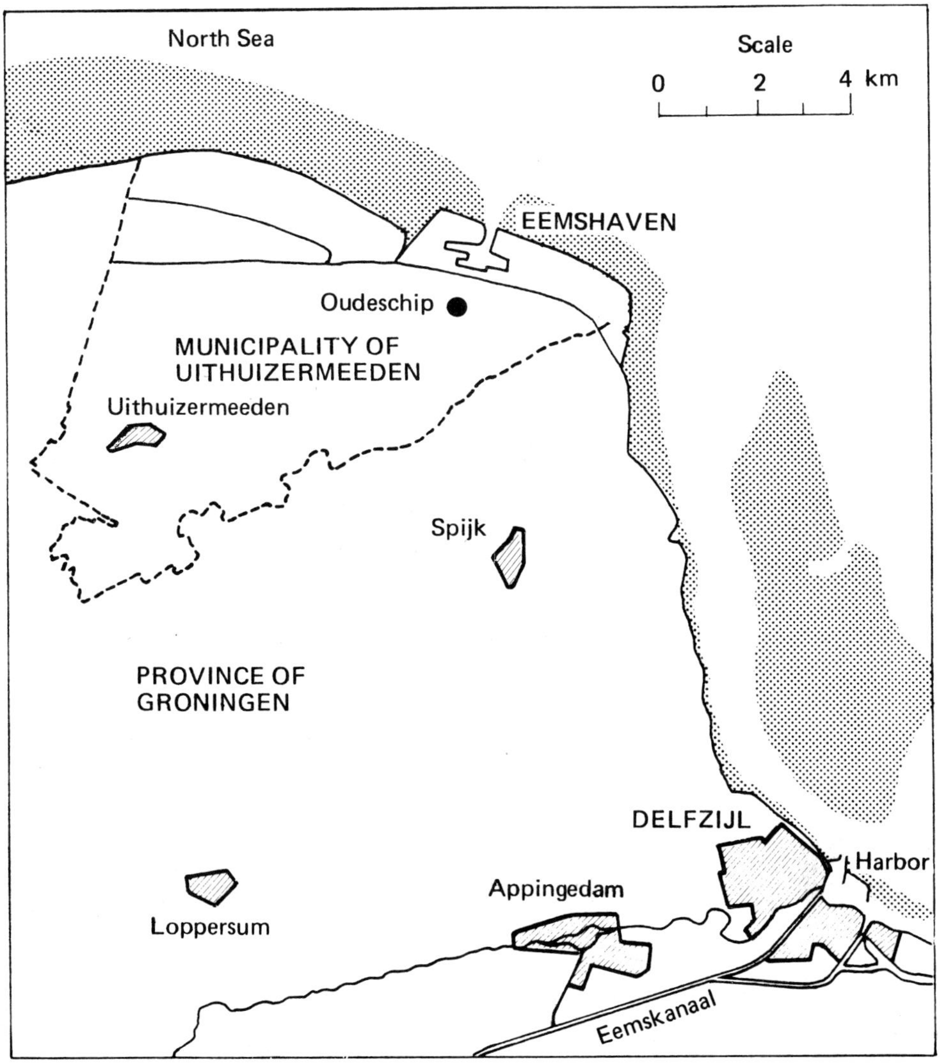

Figure 4.3. The location of Eemshaven, in the northeastern province of Groningen.

the government due mainly to their high construction costs (details of these proposals are given in Figure 4.2 and Table 4.1).

Natural gas imported by pipeline was also ruled out as a viable option at an early stage because the government believed that natural gas suppliers would be too remote to make this a feasible proposition.

SEQUENCE OF EVENTS

The decision-making process on the siting of a large-scale LNG import terminal was preceded by developments involving the government and Gasunie

Table 4.1. Details of the alternative LNG terminal sites in the Netherlands.

Maasvlakte (site A)	In southwestern corner of Maasvlakte, adjacent to existing Gasunie peakshaving plant. Relatively small site; distance to nearest towns Hoek van Holland (5 km); Oostvoorne (4 km).
Maasvlakte (site B)	In northwestern corner of Maasvlakte, larger than site A; distance to nearest towns Hoek van Holland (6 km); Oostvoorne (8 km). Shipping route to sites A and B 2 km from the center of Hoek van Holland.
Maasvlakte (site C)	Extension west of existing Maasvlakte area to be constructed. Size of area can be designed as required. Distance to nearest towns Oostvoorne (7 km); Hoek van Holland (9 km).
Voornedam Breakwater (short or long)	Extended dam to be constructed 7 or 10 km long from southwestern point of Maasvlakte. Shipping route does not interfere with other Rotterdam harbor traffic. Distance to nearest town Oostvoorne (10—13 km) (short or long dam).
North Sea island location	Artificial island to be built 20—50 km off the Dutch coast, connected by pipeline to Maasvlakte or elsewhere.
Offshore tunnel terminal system (OTTS)	Platform 4 km offshore from Maasvlakte for reception of LNG, with underwater pipeline for transport of gas to storage site at Maasvlakte. Distance to nearest town Hoek van Holland (11 km).
Eemshaven	New harbor complex at the northernmost tip of the province of Groningen. Situated in open agricultural land, with very low population density (140 people per km^2). Distance to nearest towns Oudeschip (3 km); Uithuizermeeden (6 km). Eemshaven complex is managed by the Delfzijl Harbor Authority.

in the early 1970s. Gasunie first became involved with LNG in 1972, when discussions were held with the Rotterdam Harbor Authority and local authorities concerning its plans for an LNG peakshaving plant at Maasvlakte.[3] Formal approval for the siting of such a plant, and for an LNG terminal, was the responsibility of the municipal and provincial authorities via various environmental and planning legislation (concerning pollution, nuisance, etc.).[4] In a formal sense, the approval of an LNG terminal site could similarly have been handled at local government level, provided that the Ministry of Economic Affairs had first approved the plans in accordance with economic policy.

In the case of the LNG terminal, however, government involvement went beyond that stipulated by formal (minimal) procedures; the role the government would later play in the decision process can be partly traced back to its earlier involvement in the LNG peakshaving plant decision. In addition to the involvement of the Ministry of Economic

Affairs in relation to energy policy, the government and Gasunie also discussed the safety aspects of the peakshaving plant. This led to the direct involvement of the Ministry of Social Affairs, which was concerned with occupational hazards, etc. One consequence of this latter was that a special committee was set up to investigate the safety aspects of LNG.[5]

The major interested parties involved in the LNG siting decision process included the national government (including the cabinet), Gasunie, and the local authorities of Rotterdam (for the Maasvlakte sites) and Groningen (for the Eemshaven site). In addition, many other government and some private interests influenced to varying degrees the actions and decision of these primary actors. Table 4.2 lists the most important of these groups.

The major events and decisions in the debate on LNG are summarized by a PERT diagram (Figure 4.4), representing the sequence of decisions and events. This sequence can be characterized by three main rounds (A, B, and C), each of which involved different policy questions, contextual factors, and actors. This division of the major decisions and events into rounds illustrates the main shifts that occurred during the decision process.

Round A. For our purposes the decision process concerning LNG can be said to have begun when Gasunie declared its interest in importing LNG in the early 1970s. ① Preliminary discussions with the Algerian state company Sonatrach took place in 1973 for the supply of 6 billion m^3 of LNG per year for a period of 20 years. Although a final contract was only drawn up in 1977 after a second round of discussions, the early negotiations stressed Gasunie's objective of importing approximately 10—15 billion m^3 of LNG per year by the 1990s, in line with national energy policy.[6] Through the close working relationship of Gasunie and the Ministry of Economic Affairs, the government was directly involved with the plans. ② There was also concern over the availability of a suitable and safe LNG terminal site, so that in March 1974 the Ministry of Social Affairs commissioned the Nederlandse Organisatie voor Toegepast-Natuurwetenschappelijk Onderzoek (TNO)[7] to carry out a study of the safety aspects of LNG importation, while the Netherlands Maritime Institute (NMI) examined the nautical feasibility. The government became further involved in 1975, when Gasunie officially requested the government's view of the possibility of an offshore terminal. During this time, Gasunie started discussions with harbor authorities (particularly Rotterdam) within the Netherlands, as well as abroad, and with the local authorities that would eventually be responsible for granting site approval.

In response to Gasunie's request, and existing industrial interest in the possibility of an artificial island in the North Sea, in 1975 the Ministry of Transport and Public Works set up the Stuurgroep Studie Noordzee-eilanden en Terminals (STUNET: North Sea Island and Terminal Steering

Table 4.2. Key parties in the LNG decision process in the Netherlands.

Applicant	
NV Nederlandse Gasunie:	Gasunie: the sole national gas company set up in 1963 for the management, sale, and distribution of natural gas in the Netherlands. The government holds 50% of the shares in Gasunie, participates in its governing body, and must approve or veto many of its activities.
National government	
Cabinet	The national executive body comprised of 16 ministers, responsible for making national policies and decisions (all but two head government departments).
ICONA	The Interdepartmental Coordinating Committee on North Sea Affairs (Interdepartmentale Coordinatie Commissie voor Noordzeeaangelegenheden). A policy advisory group to the cabinet comprising (civil servant) representatives of all but two of the 16 cabinet ministers.
Local authorities	
Groningen local authorities	Include (*a*) governors and council of the province of Groningen; (*b*) the city council of Uithuizermeeden; and (*c*) Delfzijl Harbor Authority.[a]
City of Rotterdam	The local authority with primary responsibility for planning permission and building permits in Rotterdam; represented by the mayor and aldermen; responsible for regulating harbor activities via the Rotterdam Harbor Authority.
Rijnmond Public Authority	A collective of 16 municipalities in the Rotterdam area, including the city of Rotterdam, that performs certain legislative roles regarding activities such as environmental planning, housing policy, transportation, health and safety, and pollution management.
Province of Zuid-Holland	Encompasses the Rotterdam area and has legislative responsibility for certain pollution, planning, and housing regulations.
Other interested parties	
Dutch Shipowners' Association	
Electricity Corporation of Groningen and Drenthe	
Provincial Chamber of Commerce of Groningen	
Public interest and environmentalist groups in Rotterdam and Eemshaven	
Trade Union organizations in Groningen	

[a] The three Groningen local authorities are considered as one body in this chapter since they held nearly the same viewpoint on the LNG siting issue and coordinated their policy actions throughout the decision process.

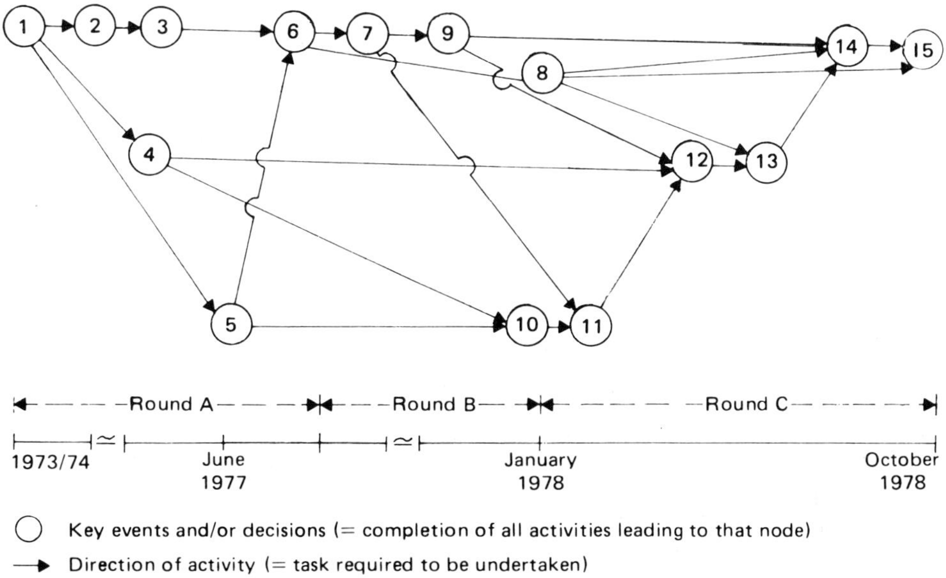

Figure 4.4. PERT diagram for the LNG decision process in the Netherlands.

Key events and/or decisions

① Gasunie declares interest in LNG imports and building a terminal (early 1970s).
② Gasunie discussions with Ministry of Economic Affairs begin (1977).
③ Government activities: TNO risk study commissioned; interest in offshore island terminal declared; STUNET and ICONA initiated.
④ Rotterdam harbor and local authorities begin study and discussion of LNG terminal siting.
⑤ Gasunie signs contract with Sonatrach for LNG supply (June 1977).
⑥ Sonatrach contract approved by Ministry of Economic Affairs (October 1977).
⑦ Cabinet and parliament start discussions on LNG.
⑧ Public, environmentalist, and other interest groups start expressing concern over LNG.
⑨ Government requests further official advice from ICONA and other relevant organizations; draws up procedure for local authority consultations.
⑩ Discussions between Gasunie and Groningen authorities start regarding Eemshaven (December 1977).
⑪ Eemshaven accepted as an alternative LNG terminal site by government (in addition to the Maasvlakte sites) (March 1978).
⑫ Local authorities start official local decision procedures (April 1978).
⑬ Local authorities formulate positions on LNG terminals at Eemshaven and Maasvlakte, respectively (June 1978).
⑭ Cabinet decision: LNG terminal at Eemshaven (August 1978).
⑮ Parliament debates cabinet decision and approves LNG terminal at Eemshaven (October 1978).

Committee), and a working group to investigate the feasibility and desirability of an offshore LNG terminal as an alternative to a land-based site. ③ The technical reports of STUNET (1977) were then submitted to a coordinating committee for North Sea affairs (ICONA).[8] STUNET advised positively *vis à vis* LNG imports and judged the "pipeline option" not advisable because of economic and practical constraints, especially the lack of "nearby" natural gas supplies (STUNET 1977). ④

A major impetus was introduced into the decision process in June 1977 when Gasunie signed the contract with Sonatrach for the purchase of 4 billion m^3 of LNG per year over the period 1985—2005. ⑤ An important deadline was thus introduced; a side-letter to the contract stipulated (*a*) approval of the contract by the Ministry of Economic Affairs by 31 October 1977; and (*b*) notification of the exact location of an LNG import terminal by 31 October 1978. If an LNG terminal site could not be announced by this date, the contract would become void.

In line with ICONA's first policy report to the cabinet, the Ministry of Economic Affairs gave official approval to the Gasunie—Sonatrach contract on 18 October 1977, although without consultation with the full cabinet (Tweede Kamer 1978). ⑥ The issue of LNG imports and the selection of a terminal site then entered more fully the political arena — involving the cabinet, parliament, local authorities, and environmentalist groups. The approval of the LNG contract signaled a new round in the decision process, to determine whether an acceptable LNG terminal site could be found within the specified time period.

Round B. With the involvement of the cabinet, and the awareness within several government departments that the siting of an LNG terminal involved a number of issues beyond energy policy (such as the environment, safety, land use, regional planning, etc.), it became clear that the final decision would have to be taken at the national level. ⑦ ICONA, the main coordinating committee on LNG, and various other advisory bodies such as the Interdepartmentale Commissie voor Milieuhygiene (ICMH: Committee for Environmental Hygiene), and the Rijks Planologische Commissie (RPC: State Land-Use Planning Committee), were therefore called in by the government to provide further advice.

The responsibility for granting final siting approval and planning permission for an LNG terminal, however, remained with the relevant local authorities. They followed established procedures in setting up public hearings and local and regional debates. The government therefore thought it necessary to request the relevant local authorities for "in advance" judgments as to the acceptability of an LNG terminal in their areas. A special decision procedure and a timetable were drawn up by the government in late 1977 for consultation with the local authorities, in order to avoid undesirable delays, and to ensure that a site selected at national level would not prove unacceptable at a later stage when the local authority would be asked to grant planning permission.

By 1977 Rotterdam harbor had become the preferred site for the terminal, following initial screening for technical feasibility and nautical safety by the NMI and other bodies. Both the government (including ICONA) and Gasunie focused virtually all their attention upon a Rotterdam location, once it had been decided that a land-based terminal was favored for strategic and economic reasons. ⑨

Initial discussions (from 1977 onwards) with the local authorities in the Rotterdam region (province of Zuid-Holland, Rijnmond Public Authority, and City of Rotterdam), indicated that they were likely to require stringent conditions before approving an LNG terminal. In particular, Rijnmond Public Authority (which had traditionally given particular attention to safety and environmental matters) indicated that it would question in detail the safety and desirability of such an operation. Safety and other local concerns thus began to endanger the tightly timed decision-making schedule. ⑧ A serious threat of delay was presented by demands from local authorities for greater public participation and for certain conditions, such as the absence of a nuclear power station in the region, before approval would be given. This potential opposition from the local authorities, in addition to new objections being raised by environmentalist groups, triggered an important new development: in December 1977 Gasunie approached Delfzijl Harbor Authority in the province of Groningen to reopen discussions as to the feasibility of using the newly built Eemshaven harbor as a site for the terminal, ⑩ thus introducing a major new group of actors — the local authorities in Groningen — and signaling a new round in the decision-making process.

Round C. Following additional feasibility and safety studies,[9] and a positive response from the Groningen local authorities, Eemshaven was seen to be a viable alternative to Maasvlakte for the LNG terminal site. In March 1978 the cabinet responded to the strong and enthusiastic requests of the Groningen authorities for official consideration of Eemshaven, and so they were included in the special decision procedure that would lead to the final site selection (Tweede Kamer 1978). ⑪

The decision-making processes at the local authority level in Groningen and Rotterdam formally began in April 1978. ⑫ They were each given three months in which to formulate their respective positions *vis à vis* the acceptability in principle of an LNG terminal in their areas, by means of council debates at various levels, and hearings where environmentalist groups and the public could air their views. These local views were presented to the cabinet in late June 1978. ⑬

The positions of the local authorities, in addition to the advice from ICONA and other advisory bodies, provided the background for the cabinet's final siting decision: in August 1978 it declared its preference for

Eemshaven, primarily on socioeconomic and regional industrial grounds (Tweede Kamer 1978). (14) The decision was debated at considerable length in parliament; it was criticized and questioned by many political parties, but was nevertheless approved in October 1978[10] (Tweede Kamer 1978/1979). (15)

These key events, especially those initiating new rounds in the decision process, emphasize the importance of the government and Gasunie (rather than the local authorities, which have responsibility for site approval only) as the two major interested parties involved in determining the direction and dynamics of the decision-making process. The local authorities in Rotterdam (particularly Rijnmond Public Authority) also played a significant role since their concerns over the safety and desirability of LNG had prompted Gasunie's investigation of Eemshaven as an alternative site. The "problem definitions" adopted by the key parties were very important in structuring the decision process, particularly in rounds B and C, in which the debate centered on the siting of the LNG import terminal, without any further questioning of the viability of other alternatives.

PARTY CONCERNS

The Dutch terminal siting decision was characterized by the involvement (either directly or indirectly) of a large number of interested parties. Although formal planning permission was to be granted at local authority level, the government, and ultimately the cabinet, took prime responsibility for the basic siting decision, moving itself to the center of the debate. Figure 4.5 shows the main actors in the approval process and the major directions of the interactions.

Table 4.3 summarizes the principal concerns of the major parties. At the national level, the government (i.e., the cabinet and interdepartmental committees) was concerned with a very wide spectrum of relevant issues; government advisors and other internal consultative bodies at civil service level (ICMH, RPC, and CPR) were requested to investigate particular aspects (e.g., safety, environmental impact, etc.), while the local authorities and Gasunie had more limited sets of concerns. Because of their different interests, the parties naturally attached varying degrees of importance to certain aspects of the decisions at hand. The analysis of party perspectives therefore reveals a set of party-specific problem formulations based on these different concerns, and the roles the parties were "assigned" in the decision process. In this section these different party perspectives are assessed, particularly those dominant dimensions[11] that seem to have governed the different party positions.[12]

ICONA. The role of ICONA was comprehensive. Utilizing studies from other official advisors and interdepartmental bodies — including those of

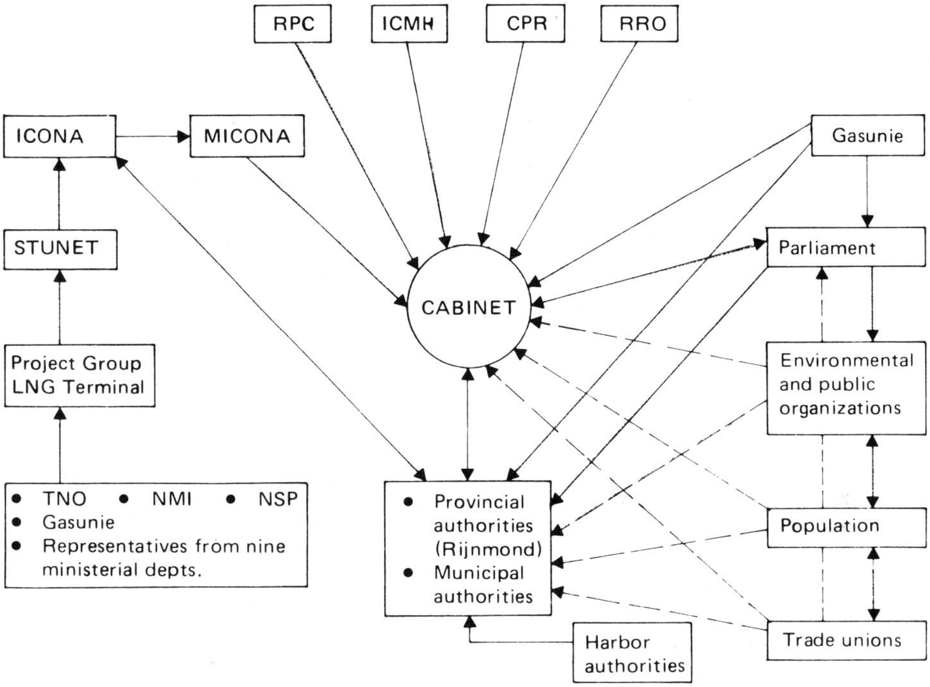

Figure 4.5. Decision-making organization and input into the Dutch cabinet regarding the LNG siting decision (based on TNO 1978).

ICONA	Interdepartmental Coordinating Committee for North Sea Affairs
STUNET	North Sea Island and Terminal Steering Committee
TNO	Organization for Applied Scientific Research
NMI	Netherlands Maritime Insitute
MICONA	Ministerial Committee for North Sea Affairs
RPC	State Land-Use Planning Committee
ICMH	Interdepartmental Committee for Environmental Hygiene
CPR	Committee for the Prevention of Disasters due to Dangerous Substances
RRO	Council for Land-Use Planning
NSP	National Shipping Laboratory

TNO, ICMH, RPC, and CPR — ICONA prepared three policy reports addressing:

- the desirability of LNG imports,
- whether the Netherlands should have its own LNG terminal;
- whether the terminal should be land-based or offshore; and
- which site was preferred and under what conditions.

ICONA was the only coordinating body that comprised representatives of all the relevant ministries and, in order to prepare a cabinet position, it took into account all aspects of national policy, as shown in Table 4.4. It is important to note that ICONA first made a detailed assessment of the

Table 4.3. Party/concern matrix.

Concerns		Parties							
		Applicant	Government					Other	
			National			Local			
		Gasunie	Cabinet	ICONA	Parliament	Groningen	Rotterdam	Environmentalist groups	Trade unions
National									
Need	N_1	x	x	x	x	x	x	x	
Import policy	N_2	x	x	x	x				
Export policy	N_3	x	x	x	x				
Economic benefits	N_4		x	x	x			x	
Economic costs:									
facility costs	N_{51}	(x)	x	x	x				
harbor costs	N_{52}	(x)	x	x	x				
channel costs	N_{53}	(x)	x	x	x				
Image	N_6		x	x	x				
Socioeconomic benefits	N_7		x	x	x			x	x
Regional									
Air quality	R_1		x	x	x	x	x	x	
Industrial development	R_2		x	x	x	x	x	x	x
Local									
Amenity and environment:									
land use	L_{11}		x	x	x	x	x	x	
fishing, marine	L_{12}		x	x	x	x		x	
tourism, recreation	L_{13}					x		x	
Economic benefits:									
taxes	L_{21}					x	x		
jobs, short-term	L_{22}		x	x	x	x	x		x
jobs, long-term	L_{23}		x	x	x	x	x		x

jobs, indirect	L_{24}		x	x	x	x	x		x
spin-off industry	L_{25}		x	x	x	x	x		x
harbor activity and prestige	L_{26}		x	x	x	x	x		x
Economic costs:									
facility	L_{31}		x	x	x	x	x		x
harbor	L_{32}		x	x	x	x	x		x
channel	L_{33}		x	x	x	x	x		x
Risks to population:									
due to shipping	L_{41}		x	x	x	x	x	x	
due to gas storage	L_{42}	(x)	x	x	x	x	x	x	
Risks to workers	L_5	(x)	x	x	x	x	x		
Consumers									
Proximity of users to facility	C_1	x	x	x	x				
Price of gas	C_2	x	x	x	x				
Supply interruption risk:									
due to need for gas	C_{31}	x	x	x	x				
due to source country	C_{32}	x	x	x	x				
due to weather	C_{33}	(x)	(x)	(x)	(x)				
due to accident	C_{34}	x	x	x	x				
due to earthquakes	C_{35}								
Applicant									
Profit:									
from revenues	A_{11}	x	x		x				
costs, facility	A_{12}	x	x		x				
costs, harbor	A_{13}	x	x		x				
costs, channel	A_{14}	x	x		x				
Control over sources	A_2	x	x		x				
Image	A_3	x	x		x				
Physical suitability of site	A_4	x	x		x				

ICONA: The Interdepartmental Coordinating Committee on North Sea Affairs. Gasunie: NV Nederlandse Gasunie
Groningen: Groningen local authorities: province of Groningen, Delfzijl Harbor Authority, and city councils in Eemshaven area
Rotterdam: Rotterdam local authorities: Rotterdam City Council, Rijnmond Public Authority, and province of Zuid-Holland
(x): Marginal concerns, not explicitly stated.

Table 4.7. City of Rotterdam policy perspective.

Policy question	Dominant dimension (s)	Outcome
LNG terminal feasible and desirable at Maasvlakte?	(1) Socioeconomics (2) Economics/cost	Yes
LNG terminal acceptable at Maasvlakte?	(1) Health and safety	Yes[a]
Preferred Maasvlakte site?	(1) Economics/cost (2) Socioeconomics	Maasvlakte site B[a]

[a] Maasvlakte would only be acceptable if no nuclear power station would be built in the area in the future.

Table 4.8. Groningen local authorities' policy perspective.

Policy question	Dominant dimension (s)	Outcome
LNG at Eemshaven acceptable?	(1) Socioeconomics (2) Health and safety	Yes
Preferred site for LNG: Eemshaven or Maasvlakte?	(1) Socioeconomics (2) Health and safety	Eemshaven

existing industrial activities in the harbor, and that it would *not* greatly increase the "cumulative" level of risk. However, one of the authorities' conditions for the acceptance of the terminal at Maasvlakte was that the (future) siting of a nuclear power plant there would be ruled out (Rotterdam 1978a, b).

Eemshaven — The Groningen local authorities. In the province of Groningen various bodies and local authorities joined forces at the start of the discussions on LNG.[14] Following Gasunie's interest in Eemshaven, the authorities were very anxious to attract the industry as a means of stimulating the economy of this relatively poorly developed northern province. In view of the perceived economic advantages that an LNG terminal would bring, particularly in terms of employment, the local authorities, as shown in Table 4.8, addressed only the policy questions concerning the conditions under which an Eemshaven terminal would be acceptable. Since Eemshaven was being introduced as an alternative to Maasvlakte, the authorities, in order to strengthen their case, undertook a direct comparative assessment of all the major sites, despite the cabinet's request that they evaluate only their own position (Groningen 1978a, b).

Gasunie. Gasunie's policy perspective was based on its obligation to supply natural gas at economical prices; its position with respect to an LNG terminal, as shown in Table 4.9, was thus related to energy policy,

Table 4.9. Gasunie policy perspective.

Policy question	Dominant dimension (s)	Outcome
Optimal site for LNG terminal?	(1) Energy policy (2) Economics/cost	Maasvlakte

and economics/cost. From both of these perspectives Maasvlakte site A had been the company's first choice.[15] In addition, Gasunie saw corporate/ strategic benefits in having an LNG terminal in the Netherlands. Although Eemshaven was also an acceptable alternative, it was only a second choice (Gasunie 1978b).

Other parties. Additional interested parties were involved in the Dutch decision-making process, but they often were less directly concerned with LNG (e.g., only local environmental or employment concerns). These parties were assigned a less institutionalized role in the debate and usually had fewer formal contacts with the central parties responsible for making the major decisions (i.e., the cabinet and local authorities).[16] The main channels of communication between these minor actors were via political parties and responsible cabinet ministers. In the Netherlands no special "anti-LNG" citizens' groups were formed, but the Eemshaven and Rijnmond cases were taken up by existing environmentalist groups through which local residents could participate. Local and national political parties also played a role in that they provided opportunities for (public) discussion via elected members of councils of representatives (municipal councils, Rijnmond council, provincial councils, and parliament).

THE MAMP FRAMEWORK

In this section the multi-attribute, multi-party (MAMP) framework,[17] which was described in Chapter 2, is applied to the case study material in order to understand better the LNG facility siting decision process in the Netherlands. Table 4.10 describes each round in terms of its problem formulation, its initiation, party interactions, and its conclusions(s). The conclusion of each round is then incorporated into the problem formulation of the subsequent round, until the final decision is made. Exogenous events, however, can enter any round at any point, and may act to reverse earlier conclusions and extend the decision-making process.

The decision process, as illustrated in Figure 4.4, was characterized by three rounds, within which the parties interacted in relation to particular problem formulations. The representation of the process in Table 4.10 through the use of rounds uncovers a set of issues that are elaborated below. These issues highlight (1) the introduction and rejection of alternative

Table 4.10. The MAMP framework applied to the decision process on LNG in the Netherlands.

(*a*) ROUND A: December 1976—late 1977.

I PROBLEM FORMULATION

Assumptions:	(1) A government policy paper (1974) has established the *need* to import LNG in order to conserve Dutch natural gas supplies. (2) Gasunie is interested in locating an LNG terminal adjacent to its peakshaving plant at Maasvlakte (Rotterdam harbor). (3) Government and industry are jointly interested in developing an artificial island in the North Sea.
Questions:	(1) Is it feasible and timely to have an LNG terminal? (2) If so, should the plant be Dutch or foreign?, offshore or land-based?, located at Maasvlakte?
Process constraints:	(1) Any LNG import contract must be approved by the Ministry of Economic Affairs (MEA). (2) Siting approval would proceed through local authorities.

II INITIATION

Gasunie declares its interest in importing LNG, ① and requests the formal approval of the Ministry of Economic Affairs. ② Gasunie also requests the government's official view of the acceptability of a terminal at Maasvlakte. ③④

III INTERACTION

Party	Position	Arguments
Gasunie	For importation of LNG to Rotterdam (Maasvlakte)	Economical and practical; existing infrastructure; technically feasible;[a] proximity of major gas users.
STUNET (1977)	LNG feasible and practical; land-based site more cost-effective; island terminal preferable on safety grounds	Pipeline not viable and land-based site conditions less costly; Eemshaven not acceptable on nautical grounds; *risk comparable to other industrial risks*; environmental effects acceptable.
ICONA	LNG import desirable; Dutch terminal (Maasvlakte, if land-based)	Diversification of energy supply, environmental benefits of gas, independence of supply, cost considerations land-based; *risk acceptable*;[b] nautical/technical grounds favor Maasvlakte (of land-based options).
Rotterdam harbor	Site at Maasvlakte	Socioeconomic benefits

IV CONCLUSIONS

(1) June 1977: Gasunie signs contract with Sonatrach (Algeria) for LNG supply. ⑤ This contract specifies that an LNG site be chosen by October 1978.

(2) October 1977: MEA approves Gasunie—Sonatrach contract. ⑥

(3) Late 1977: Government formulates a formal decision procedure for site approval, involving early consultation with local authorities.

Italics indicate arguments involving population risk. Circled numbers refer to nodes in the PERT diagram in Figure 4.4.

[a] Rotterdam (1977).

[b] TNO (1976).

Table 4.10 (*continued*). The MAMP framework applied to the decision process on LNG in the Netherlands.

(*b*) ROUND B: February 1977—March 1978.

I PROBLEM FORMULATION

Assumptions:	(1) Official government view is that the Netherlands needs to import LNG; preferred site is Rotterdam harbor. (2) The Algerian LNG contract requires a decision on a site by October 1978.
Questions:	(1) Should a site in Rotterdam harbor be approved? (2) If so, which of five possible sites: Maasvlakte (A, B, C), Voornedam, or an offshore island?
Process constraints:	The government has established a decision procedure for site approval involving consultation with local authorities.
Outcome constraint:	Government advisory bodies are strongly in favor of a Dutch site.

II INITIATION

Gasunie applies to Rotterdam for approval of an LNG terminal at Maasvlakte. The Ministry of Economic Affairs requires local authorities (at Maasvlakte) to start decision proceedings (October 1977). ⑦⑧⑨

III INTERACTION

Party	Position[a]	Arguments
City of Rotterdam	For Maasvlakte site; prefer A or B to C[b]	National need; increased harbor activity; jobs; shipbuilding; costs.
Rijnmond Public Authority	For Rotterdam site for storage; against A, B except for storage; receiving terminal should be offshore	National need; increased harbor activity; close to users; more jobs; *high transport and handling risks.*[c]
Zuid-Holland	For Rotterdam site; A, B unacceptable; for Voornedam	National need; economic factors; *risk unacceptable.*
ICONA	For Maasvlakte sites A, B, C; prefer to offshore site	National need; cost considerations; *risk acceptable.*
Ministry of H&E	Opposed to Maasvlakte site	National need questionable; *high risks.*

IV CONCLUSIONS

(1) No decision was taken on Rotterdam harbor since a new round of discussions had begun comparing the Rotterdam sites with Eemshaven. ⑩

(2) Government decides LNG site options are now limited to two areas: Maasvlakte (A or B) and Eemshaven.

[a] *Full* party positions of Maasvlakte local authorities were not officially formulated until round C had commenced.

[b] The acceptance of LNG at Maasvlakte was subsequently made *conditional* on the absence of a (future) nuclear power plant in the area.

[c] TNO (1976).

Table 4.10 *(continued)*. The MAMP framework applied to the decision process on LNG in the Netherlands.

(*c*) ROUND C: March 1978–November 1978.

I PROBLEM FORMULATION

Assumptions: (1) Official government view is that the Netherlands needs to import LNG.
(2) The LNG contract requires a decision on a site by October 1978.
(3) There are growing concerns about public acceptance and local authority approval of a Maasvlakte site.
(4) Eemshaven harbor, contrary to earlier reports, appears to be nautically acceptable for LNG, and becomes second contender, in addition to Maasvlakte sites.

Question: Should Maasvlakte (site A or B) or Eemshaven be approved for the LNG terminal?

II INITIATION
(1) Because of growing concerns about local approval of Maasvlakte, Gasunie approaches Eemshaven. ⑩
(2) Cabinet responds positively to the province of Groningen requests to take Eemshaven into consideration (March 1978). ⑪ ⑫ ⑬

III INTERACTION

Party	Position	Arguments
Groningen	For Eemshaven	Socioeconomics; employment; *risks acceptable.*[a]
Trade unions	For Eemshaven	Socioeconomics.
Environmentalist groups	Against Eemshaven	*Risks unacceptable*;[a] alternative options, such as pipeline, neglected.
ICONA	Prefers Maasvlakte to Eemshaven	Proximity to major gas users; coal gasification opportunities; costs, indirect employment; *risks approximately equivalent.*
Shipowners' Association	For Maasvlakte	*Nautical disadvantages and risks at Eemshaven.*
Electricity Corporation	Against Eemshaven	*Safety of existing coal-fired power station threatened.*
Cabinet	For Eemshaven	Socioeconomics; regional policy; *risks acceptable.*
Parliament (majority)	For Eemshaven	In line with cabinet.

IV CONCLUSIONS
(1) Cabinet selects Eemshaven; ⑭ decision approved by parliamentary majority.[b] ⑮
(2) Groningen authorities (local and provincial) indicate that they will conditionally approve the Eemshaven site.

[a] TNO (1978).
[b] Some parliamentary parties (minority) opposed to LNG siting decision made by cabinet. Arguments: risks uncertain, or unacceptable; alternative options unexplored.

LNG terminal sites (with perceived viability); (2) shifts in the decision process and their determinants; and (3) the variable influence of decision events on the final outcome.

(1) During round A, Maasvlakte emerged as the only viable land-based site in terms of nautical feasibility and safety. This conclusion was drawn at national level by ICONA, based on technical advice from STUNET and the NMI. During round A and most of round B this view was not challenged by any of the interested parties. This acquiescence emphasizes the role of the government, and perhaps to a lesser degree Gasunie, in establishing the terminal location options to be considered.[18] At an early stage of round A the government had rejected the alternative of importing natural gas by pipeline. This initial policy decision formed a particular problem formulation that guided the later decision procedure.

The introduction of Eemshaven as an alternative — following new technical studies commissioned by Gasunie and the Groningen local authorities — was made independently of the government, signaling a reduction in its influence. Under pressure from the Groningen authorities, and with the threat of local opposition to a Maasvlakte site, the government was forced to acknowledge the re-emergence of Eemshaven into the debate, and to include evaluation of this site in its formal decision procedure.

(2) The declining influence of the government, as distinct from its formal responsibility, can also be traced in the main shifts in the decision process. Following the initial decision to build a terminal, the government played an important role in initiating risk studies and setting up interdepartmental study groups and advisory bodies such as ICONA and STUNET. But two significant events outside the government's control caused its role to become more responsive, and thus brought about shifts in the dynamics and problem bounds of the debate.

The first shift occurred as a result of the LNG import contract with Sonatrach, thereby introducing round B. This required a terminal site to be selected within 16 months (by October 1978), thereby imposing a strict time limit on the proceedings. Regarding the contract, the government had relatively little influence on the activities of Gasunie, since the latter was operating as a member of a consortium involving the FRG firms Ruhrgas AG and Salzgitter Ferngas GmbH who were leading the negotiations with Sonatrach. At a broader level, Gasunie had been given a mandate to import LNG, in line with national energy policy, so that approval of the contract by the Ministry of Economic Affairs could be seen as merely a formality. The government was faced with a strong imperative to select a terminal site within the time stipulated, and so felt the need to set up a special decision procedure for "formal" consultations with the local authorities (at Rotterdam!) rather than following the normal, potentially lengthy, procedure for site approval at the local level, which might jeopardize the contract. The structure and dynamics of the decision process were largely determined by the "frame of refererence" provided by national energy policy considerations.

The second major shift in the concluding stages of the decision process was Gasunie's introduction of Eemshaven as an alternative to Maasvlakte. Gasunie appoached the Groningen local authorities to investigate the feasibility of building a terminal at Eemshaven because of the lack of consensus on the acceptability of the Maasvlakte sites based on one overriding factor — safety. Although few formal discussions had taken place at the local authority level at the time Gasunie began pursuing alternatives (December 1977), it was clear that opposition to LNG was present within the Rijnmond Public Authority, Rotterdam City Council, and the Zuid-Holland Provincial Council. Reports in the media indicated that the main issue concerned the potentially adverse environmental impact, particularly the increased safety risks an LNG terminal would pose to the local population. In the absence of agreed criteria for defining "factual risk" (probability × consequence, as used in the TNO study), concerns were also aired about the "perceived" risk level to local residents. Most of the discussions on safety of the Rotterdam sites centered on the *consequences* of a major accident.[19] The government therefore had to include Eemshaven as an alternative *in response* to a request from the Groningen authorities.

(3) Whatever the real ability of actors at the national level to influence the decision process, the final approval of Eemshaven was the responsibility of the cabinet, which placed emphasis on the dimensions of regional development policy and socioeconomic factors such as employment (see Table 4.11; Tweede Kamer 1978).[20] The overriding importance attached to these factors is emphasized by the fact that the cabinet rejected the recommendations of its closest advisors and diverged from earlier government statements on the issue (in line with national energy policy and, to a lesser extent, economics/cost, ICONA preferred Maasvlakte).

With regard to safety, the cabinet concluded that "no clear preference" could be given to either site, but it did not resolve the conflicts that were apparent between interested parties on the basic issues of desirability or acceptability of LNG. The government did not involve itself in detailed discussions with opponents of the terminal (such as environmentalist groups); nor did it *specifically* endorse the view of the Groningen authorities and others that the risks involved at Eemshaven or Maasvlakte were equally acceptable.[21] The cabinet did not actually use safety as a final selection criterion, and so did not make any fundamental objections to LNG on grounds of safety, but focused instead on dimensions such as socioeconomics.[22]

The above course of events suggests that the outcome of the LNG decision process cannot be analyzed solely in terms of the party interactions represented in the MAMP framework. In particular, the clearly divergent views of the cabinet and its many official advisory bodies (especially ICONA), on factors such as the "psychological" importance of

Table 4.11. Government policy perspective.

Policy question	Dominant dimension (s)	Outcome
LNG imports desirable?	(1) Energy policy	Yes
Dutch terminal desirable?	(1) Energy policy	Yes
Land-based or offshore terminal?	(1) Economics/cost (2) Energy policy	Land-based
Preferred LNG site?	(1) Socioeconomics	Eemshaven

siting the terminal at Eemshaven to provide a stimulus to industrial development, limits the extent to which the cabinet's final decision can be understood in terms of the official advice submitted to it.

Indeed, there are indications that a significant factor was political pressure: the Groningen authorities had enlisted support at the provincial and municipal levels, as well as from chambers of commerce and trade unions, and were thus able to present a united front. In Rotterdam, on the other hand, the authorities were divided: the City of Rotterdam governors were strongly in favor of a Maasvlakte site for the terminal, but the Rijnmond Public Authority was reluctant, if not unable to give its wholehearted support. Another political factor regarding the acceptability of Maasvlakte was the condition made by the Rotterdam City Council that a nuclear power plant would not be sited in the area in the future.

The Groningen local authorities argued that a decision in favor of an LNG terminal at Eemshaven would provide a unique opportunity for the government to endorse, through concrete actions, its stated policy of encouraging industrial development in the northern provinces. The Eemshaven harbor complex, which had been developed with government support, had by and large remained unused since its official opening in 1973. This argument was endorsed by the governor of Groningen, a skillful politician with experience at cabinet level. He was in a position to penetrate cabinet circles through personal contacts and to exploit his ties with one of the political parties (VVD) in the coalition government, of which he was a longstanding member.[23] Without attaching perhaps too great an importance to the influence of individuals, the role played by the governor of Groningen does indicate the importance of factors clearly outside the formal decision process that may have increased the weight of some arguments put forward by interested parties. The final decision therefore appears to have been guided more by political opportunity than by consistent government policies, strategies, or decision procedures. The final outcome conflicted with the official advice of ICONA, and raises the issue of the appropriateness and effectiveness of "expert advice" in this decision process. ICONA (purposely) left out of consideration two important political aspects: the local risk perception in relation to public and official acceptance of LNG, and the political importance attached by some interested parties to the siting of the terminal. In retrospect, these factors can be seen to have been the most dominant in influencing the structure and outcome of the siting decision process. It should be emphasized that the nature of the "decision problem" was framed by the government in terms of energy policy in general, and the perceived need for liquefied gas imports in particular.

THE RISK ISSUE

Risk and safety were important considerations in the Dutch decision-making process on LNG. The government's early concern was demonstrated in 1974, when it commissioned an LNG risk assessment by the TNO. The TNO report (1976) considered the risks involved in an LNG terminal at Maasvlakte, and at the request of Groningen local authorities, the analysis was later revised for a terminal at Eemshaven (TNO 1978). No other risk studies were undertaken, and all major parties made use of the TNO assessments. These were technical documents that excluded specific discussion of *acceptable* levels of risk. They did not consider the "perceived risk", but used the common concept of risk: "probability × consequence". In this context it is relevant to review the positions of the various interested parties in the LNG debate in relation to the qualitative assessment of risk, as shown in Table 4.12.

Confronted by the various positions on the risk of LNG, the cabinet decided that both Maasvlakte and Eemshaven were acceptable sites, but failed to state specifically what analysis formed the basis of this conclusion.[24] It stated only that the risks were acceptable in relation to the perceived advantages of importing LNG. As regards the option of an off-shore terminal (as was preferred by some groups, such as the Rijnmond Public Authority), the cabinet decided that the disadvantages, in terms of costs and timing, outweighed the marginal safety advantages. In the recognition of such agreed acceptable levels of risk, safety aspects were only discussed in relative terms: (a) by comparing the risk of LNG with existing risk levels of other activities, and (b) by shifting the discussion towards an assessment of the comparative advantages of different sites.

It is relevant to distinguish between the impact of safety concerns upon the structure of decision *process*, and the *outcome* of the final round of the debate. The risk dimension was very important in the structure of the decision process and the nature and direction of the various rounds. In this context the following aspects may be noted:

(1) Concern about safety aspects of LNG by the government was expressed at a very early stage[25] and led to the commissioning of the TNO risk assessments.
(2) Nautical safety was an important consideration in the initial screening

Table 4.12. Positions of interested parties regarding risk in the Dutch LNG debate, based on TNO (1980) and Schwarz (1982).

Acceptable	Too uncertain	Additional population risk unacceptable	Unacceptable
Cabinet	Zuid-Holland Provincial Council	Zuid-Holland Provincial Governors	North Sea (Noordzee) Environmentalist Group
ICONA	Eemsmond Environmentalist Group	Rijnmond Public Authority	Electricity Corporation of Groningen
Parliament (majority)	North Sea (Noordzee) Environmentalist Group	Minister of Health and Environmental Hygiene (minority view)	
Rotterdam Harbor Authority			
Rotterdam municipal authorities			
Groningen local authorities			
RPC			

Notes
Acceptable: risks are negligibly small or acceptable in relation to the advantages of LNG.
Too uncertain: the risk analyses are too uncertain; too many underlying assumptions and contradictions; it is unacceptable to draw conclusions (at this stage); further investigation of risk and alternative options should be pursued.
Additional population risk unacceptable: psychological factor/perceptions of risk; at least handling/reception of LNG should not take place at Maasvlakte (parties in this group did not express views on the acceptability of the risk at other locations, nor on absolute levels of acceptable risk).
Unacceptable: possible consequences of an accident are too great; reception/handling and storage of LNG onshore are unacceptable.

of alternative LNG sites, and led to the preference for a Maasvlakte site (setting the context for round A).

(3) Following the signing of the contract between Gasunie and Sonatrach (introducing round B), discussions about safety intensified at many levels, involving the Rotterdam authorities, environmental and public interest groups, parliament, and political parties.

(4) Approval of the LNG contract by the Ministry of Economic Affairs took place after the ICONA assessment had concluded that the risks involved in a land-based or offshore LNG terminal would be equally acceptable.

(5) Gasunie's introduction of Eemshaven as an alternative site reflected concern about the timing of the approval by the Rotterdam authorities, which in turn was the result of discussions on the local opposition

to the safety risks involved, especially the potential consequences of a major accident (prerequisites to round C).

(6) The subsequent (re-)introduction of Eemshaven into the debate (signaling round C) took place only after re-examination and approval of the nautical and safety aspects involved at Eemshaven harbor by the Groningen local authorities.[26]

However, in relation to the final outcome in favor of Eemshaven, the risk dimension was relatively insignificant. The final choice of a site had been narrowed to Maasvlakte and Eemshaven, and the cabinet made its decision on the assumption that it was practically impossible to obtain agreement on general safety criteria with which to assess the acceptability of an LNG terminal site. Also, ICONA and others had recommended that the risks from a land-based terminal would be acceptable.

The final choice of Eemshaven was justified by the cabinet primarily in terms of socioeconomics, particularly the positive effects of an LNG terminal on regional development and employment. The fact that the government's own risk assessment showed that the Eemshaven site was in some respects safer for the local population than Maasvlakte, seems *not* to have been a decisive criterion in the final selection (Table 4.13). In fact, the cabinet specifically stated that "on the grounds of risk considerations, no clear preference can be given in favor of one or the other harbor site" (Tweede Kamer 1978). Although the government committed itself to an LNG terminal site, responding to an "energy policy" imperative, *political* considerations rather than safety concerns most strongly influenced the final outcome of the decision process.[27]

CONCLUDING REMARKS

The following concluding observations are noted in relation to the Dutch decision process.

- The decision process in the Netherlands was characterized by conflicting interests and perceptions of various interested parties, concerning in particular: (i) the need to import natural gas in liquefied form; (ii) the acceptability of the "factual" risks (probability × consequence), as well as "perceived" risks involved in LNG (e.g., by focusing on the maximum possible consequences).
- The conflicts arising from the different party concerns and problem definitions were never resolved by the government; the final outcome of the decision process cannot be understood solely in terms of the party-specific interactions.
- The final selection of the LNG terminal site was a *political* decision, influenced by the government's commitment to a national energy policy favoring imports of LNG to a land-based terminal. The site had

Table 4.13. Risk comparisons of Maasvlakte and Eemshaven (from Tweede Kamer 1978).

	Maasvlakte	Eemshaven
Probability of major accident (after additional safety measures)	3×10^{-7} (3×10^{-8})	10^{-7} (5×10^{-8})
Maximum consequences		
Number of deaths	$0.5–2 \times 10^4$	$0.5–2 \times 10^3$
Number of casualties	$1–4 \times 10^4$	$1–4 \times 10^3$
Material third-party damage (Dutch guilders)	18×10^9	?
Increase in risk of individual death	3×10^{-6}	$< 3 \times 10^{-7}$
Weighted risk after safety measures	0.028	0.023 (approx.)

to be selected within a specific time period to enable the Algerian LNG supply contract to be honored. The cabinet decision was furthermore founded on the contention that the benefits would justify the risks involved, and that factors relating to the "political feasibility" (or acceptance) of a particular LNG terminal site was an important element in determining the final decision outcome.

From a broader perspective, it is significant that the dynamics and the "problem definition" governing the decision process on LNG siting were largely determined during round A. In the early stages, the proceedings mainly involved the applicant (Gasunie) and a limited number of government departments concerned with the planning of national energy policy. The initial accessibility of various other interested parties to the decision-making process was significantly constrained. This had implications both in terms of setting the political agenda and in the types of "legitimate" policy options that were incorporated into the decision process. Policy "demands" by local authorities and other groups upon the government were limited mainly to the later stages of the formal decision process. By the time the process entered round B, the imperative to honor the basic government commitment to importing LNG to a land-based terminal had already been established. Both the mode of public discussion on the safety aspects of LNG, and the role assigned to formal risk analyses in the LNG decision-making process in the Netherlands, must be viewed in this particular context.

NOTES

1. The shares of Gasunie are divided as follows: the state, 10%; DSM Aardgas BV (DSM, Dutch State Mines, is a state-owned company), 40%, Shell Nederland BV, 25%; and Esso Holding Company Holland, 25% (Gasunie 1978a).

2. State involvement in the governing body of Gasunie takes place through representatives of the Ministry of Economic Affairs (one of 16 cabinet ministers). The minister also has to approve Gasunie decisions concerning the annual sales plan, gas prices, and the construction of facilities and equipment for the transport and storage of gas (Tweede Kamer 1974).

3. See Chapter 1 (p2) for a discussion of LNG peakshaving plants.

4. In the case of Rotterdam an additional intermediate body was also involved: the Rijnmond Public Authority, effectively operating between the provincial and municipal authority levels with respect to certain areas of policy (see Table 4.2 for details).

5. In the context of the LNG decision-making process, it is of relevance here to mention that Maasvlakte was selected by Gasunie for its peakshaving plant on economic grounds and because of the proximity of major gas users, but potential expansion of the site for an LNG terminal at a later date had also been considered. In the early 1970s, as Gasunie was planning the peakshaving plant, the company was already involved in discussions with Algeria regarding imports of LNG, for which a terminal would eventually have to be built. The peakshaving plant was approved in the mid-1970s by the local authorities of Rotterdam (after considerable discussion on safety and other risks) and became operational in May 1977. For a brief review of the general use of risk analyses in the Netherlands to date, see Blokker (1981).

6. Discussions with Sonatrach took place within a consortium involving the FRG firms Ruhrgas AG and Salzgitter Ferngas GmbH. The annual gas consumption in the Netherlands is about 44 billion m^3, less than half of the domestic gas production (CBS 1979). The other half is exported, mainly to Italy and France, under long-term contracts made in the mid–late 1960s for 20–25 year periods. The energy situation today is different, however. Since a large part of the energy infrastructure is designed to use natural gas, Dutch gas fields are being conserved, and so the government and Gasunie agreed to embark on a policy of *importing* LNG to offset the exports.

7. TNO is a government-supported institute for applied scientific research. The TNO risk analysis *Evaluatie van de gevaren verbonden aan aanvoer, overslag en opslag van vloeibaar aardgas* (TNO Bureau Explosieveiligheid) was first published in 1976.

8. ICONA (Interdepartmentale Coordinatie Commissie voor Noordzee-aangelegenheden) included representatives of 14 government ministries, and was designed to give policy advice on North Sea affairs, including LNG, directly to the cabinet. Officially, ICONA advised MICONA, a cabinet subcommittee consisting of ministers of the

departments represented in ICONA (at senior civil servant level). Because of the way the LNG siting issue had entered at the level of national government, as a multidimensional problem, ICONA became the central forum for interdepartmental discussion on siting policy (see ICONA 1978c).

9. New nautical studies concluded that since 1976, when the NMI had first investigated the possibility of an LNG terminal at Eemshaven, changes had occurred in the approach to the port, making it (under certain conditions) feasible as an LNG harbor. TNO was therefore commissioned to carry out a risk analysis of the site (Tweede Kamer 1978).

10. A major criticism of some left-of-centre political parties (and indeed some environmentalist groups) concerned the lack of serious attention and analyses devoted by government to the option of importing natural gas via pipeline (and/or rearranging deals with major *importers* of Dutch natural gas).

11. The dimensions are the main aspects of the policy problem that were recognized and addressed by the different interested parties. The following main dimensions are used: energy policy (N_2, N_3, N_{11}, A_2); economics/cost (L_3, N_5, A_1); health/safety (L_4, L_5); socioeconomics (N_7, L_2, L_3); and environmental impact (R_1, L_1). A discussion of these dimensions and further details of the various party perspectives are given in Schwarz (1982).

12. Mention of one or more dimensions as being dominant in policy questions does not mean that other concerns were left out of consideration.

13. The City of Rotterdam favored Maasvlakte B (over C) because of cost considerations, and (over A) because of the scope for additional related industrial activities.

14. This alliance included the provincial authority, Delfzijl Harbor Authority, and the municipality of Uithuizermeeden within which the proposed terminal site was located.

15. Maasvlakte A was situated adjacent to Gasunie's LNG peakshaving plant, and therefore offered economic and operational advantages.

16. The following "minor" parties were also involved in the decision process:
    - *Trade unions*: in favor of an LNG terminal in Eemshaven. Dominant dimension, socioeconomics.
    - *Environmentalist groups*: opposed the location of an LNG terminal at Eemshaven and to a lesser degree at Maasvlakte. Dominant

dimensions, environmental impact and health; they also questioned the need to import *liquefied* natural gas.

- *Dutch Shipowners' Association*: opposed the location of an LNG terminal at Eemshaven on grounds of nautical safety.
- *Electricity Corporation* (Groningen and Drenthe provinces): opposed the location of an LNG terminal at Eemshaven since it would endanger a nearby electricity power station if there was a serious accident. For a detailed discussion of these and other party perspectives, see Schwarz (1982).

17. For further details of this framework, see Kunreuther *et al.* (1982).

18. In the event, the conclusions of government-sponsored studies of the most suitable sites were in line with Gasunie's preference for siting the terminal at Maasvlakte.

19. The number of deaths in the Maasvlakte area, in the event of a worst possible LNG accident (involving detonation), had been estimated at 5500 average, and 17 600 maximum (TNO 1976).

20. Table 4.11 reflects those dimensions that were instrumental in resolving the various policy questions. Contrary to what the table may suggest, the government did make an extensive assessment of the safety aspects of LNG.

21. The selection of an LNG terminal site was the first major environmental issue faced by the newly established Rijnmond Public Authority; given its special interest in environmental protection and the absence of legislation covering "environmental impact statements", its institutional interest was obvious.

22. Also, while the pipeline option was in all probability less risky than importing LNG, it was claimed by some that this alternative was never adequately evaluated by the government.

23. The governor of Groningen (Commissaris der Koningin) at the time was Mr E.H. Toxopeus. If the dimensions of socioeconomics, energy policy, and health and safety were crucial in the discussion on LNG siting, it is perhaps significant that the responsible cabinet ministers in those areas were members of the VVD (Volkspartij voor Vrijheid en Democratie, People's Party for Freedom and Democracy, a conservative–liberal party), with whom the governor probably had close political ties. It should also be noted that prior to December 1977 the cabinet did not include members of the VVD, but was a coalition including the Labor Party (PvdA). Local authorities concerned with approval of a Maasvlakte site had, by and large, Labor party majorities, and this is likely to have provided more scope for political contact with the cabinet prior to December 1977. The Labor Party was on the

whole more critical of the government's plans to site an LNG terminal in the Netherlands, particularly on safety grounds and the lack of attention paid to alternative gas import methods.

24. It should be noted that the various local authorities had also discussed LNG siting without explicit reference to an absolute qualitative level of acceptable risk.

25. In 1974, the need for a Dutch LNG terminal was a theoretical one, since no contracts had been drawn up.

26. The Groningen local authorities commissioned safety studies by the TNO and the provincial public works' authorities of Groningen.

27. It must be noted, however, that the cabinet was probably reluctant (for political reasons) to emphasize the safety advantages of Eemshaven as a major justification for its policy choice, given that in the early decision rounds the government had consistently played down the safety issues as a critical factor in the selection of an LNG site.

Chapter 5

The UK: Sparks at Mossmorran–Braefoot Bay*

This chapter provides a review of some aspects of the decision and approval process involved in the siting of liquefied energy gas (LEG) facilities at Mossmorran and Braefoot Bay in Fife, Scotland. The terms of reference for this decision process were for the international oil companies Shell and Esso to obtain outline planning permission (i.e., official approval in principle) for the following:

(*a*) An application by Shell for natural gas liquids separation facilities at Mossmorran, and associated jetty facilities at Braefoot Bay.
(*b*) An application by Esso for an ethylene cracker plant at Mossmorran and associated jetty facilities at Braefoot Bay.
(*c*) An application by Esso for industrial development at Mossmorran.

This study will be treated essentially as a one-site decision process. The overriding question was whether Mossmorran—Braefoot Bay, in principle, would be a suitable site for the facilities proposed. A related question, by way of qualifying any positive response, concerned the nature of the planning conditions that would need to be stipulated if this site were to be used for the LEG facilities proposed. In the case of a negative response to the first question, an alternative site would be sought and a corresponding, but essentially separate, decision process would be set in train. Alternative sites would thus be considered sequentially rather than simultaneously. The decision process spanned three years, from July 1976 when initial interest in the sites was expressed, until August 1979, when an official decision of approval was announced.

*This chapter was written by Sally Macgill, based on a more comprehensive case study (Macgill 1982). The author worked closely with Joanne Linnerooth in integrating the case study material within the MAMP framework.

SETTING THE STAGE

The facilities at Mossmorran and Braefoot Bay were planned in connection with the exploitation of the large Brent oil and gas field in the UK sector of the North Sea (see Figure 5.1). Although the main resource of the Brent field is crude oil, there are also commercially exploitable quantities of gas (methane and natural gas liquids). The installations as outlined in application (*a*) were required in order to produce the commercial fuels

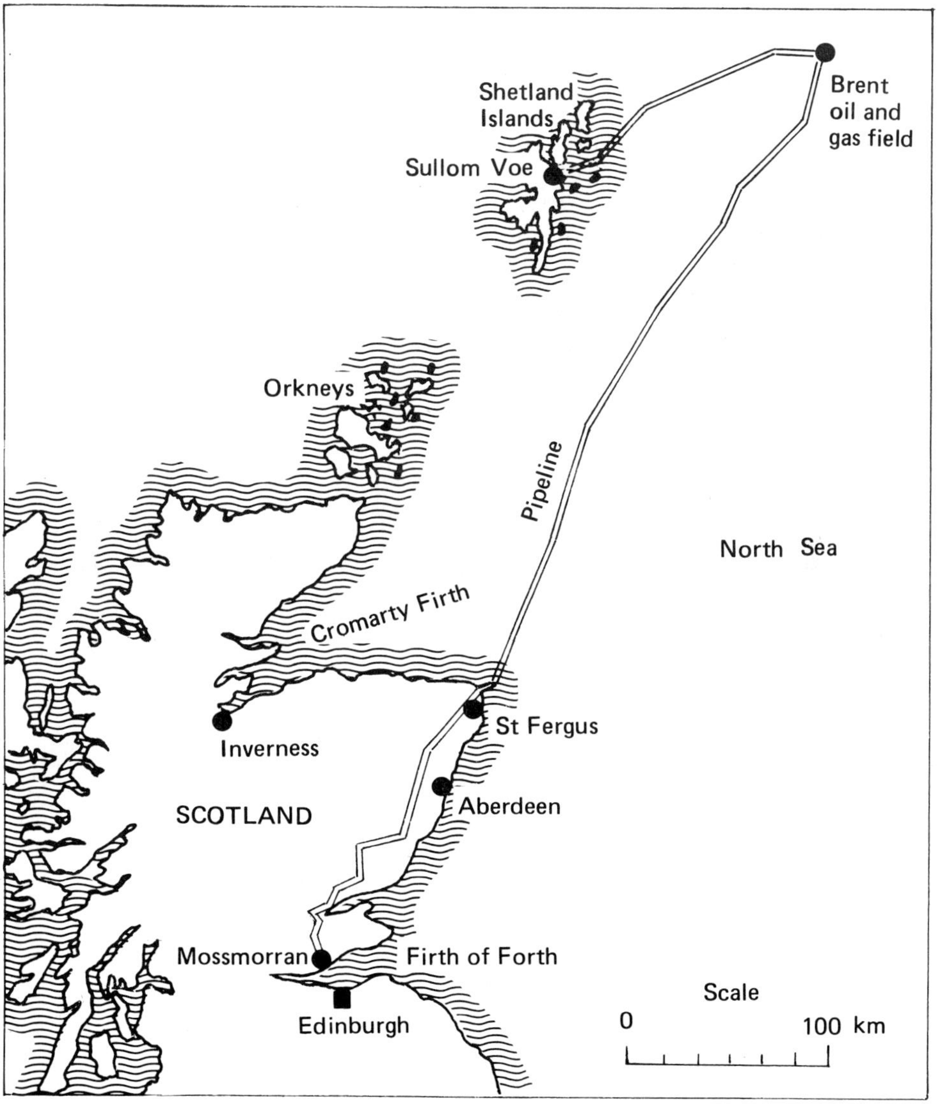

Figure 5.1. Scotland and the Brent oil and gas field.

and feedstocks (propane, butane, and ethane) from the natural gas liquids. The efficient exploitation of the North Sea reserves and their downstream processing in the UK are important aspects of UK energy and industrial policy. The commercial fuels produced at Mossmorran were intended for export through the Braefoot Bay shipping terminal (see Figures 5.2 and 5.3).

Application (*a*) was the only one of the three that could strictly be called an LEG facility, but it was considered alongside the larger package of petrochemical developments involving applications (*b*) and (*c*). The intended location of these facilities and their capacities and interrelationships are indicated in Figures 5.3 and 5.4. The proximity of the shipping terminal site at Braefoot Bay to existing communities, notably Aberdour and Dalgety Bay, was the cause of determined and prolonged opposition to the applications on grounds of safety.

During the decision process, application (*a*) was the only one for which there was a stated commitment to construct the facility in the event of permission being granted. Firm commitment by Esso to construct the ethylene cracker (application *b*) was not given until October 1980, and to date there has been no firm commitment to take up application (*c*). Evaluation of the overall benefits from the three applications was severely frustrated by this lack of commitment, as application (*c*), providing potentially

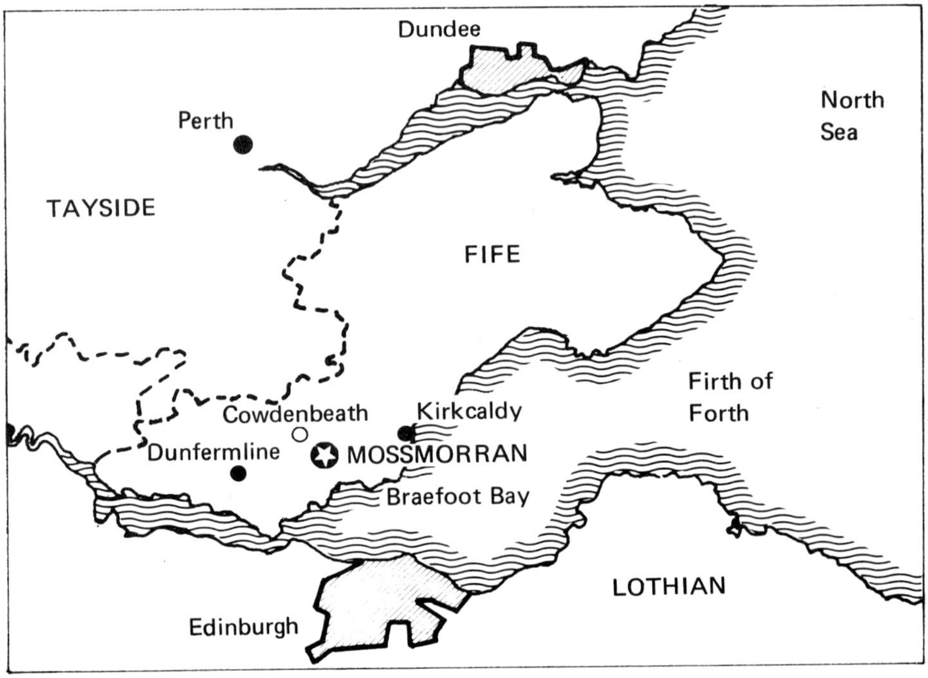

Figure 5.2. The Fife area.

in excess of 1000 permanent jobs, would be considerably more labor-intensive than either (*a*) or (*b*), which would provide only 100 and 250 jobs, respectively.

PERTINENT EVENTS

The UK has a very comprehensive planning system. Under the Town and Country Planning (Scotland) Act 1972, any development of land (whether for a householder intending to build a garage adjacent to his house, or a

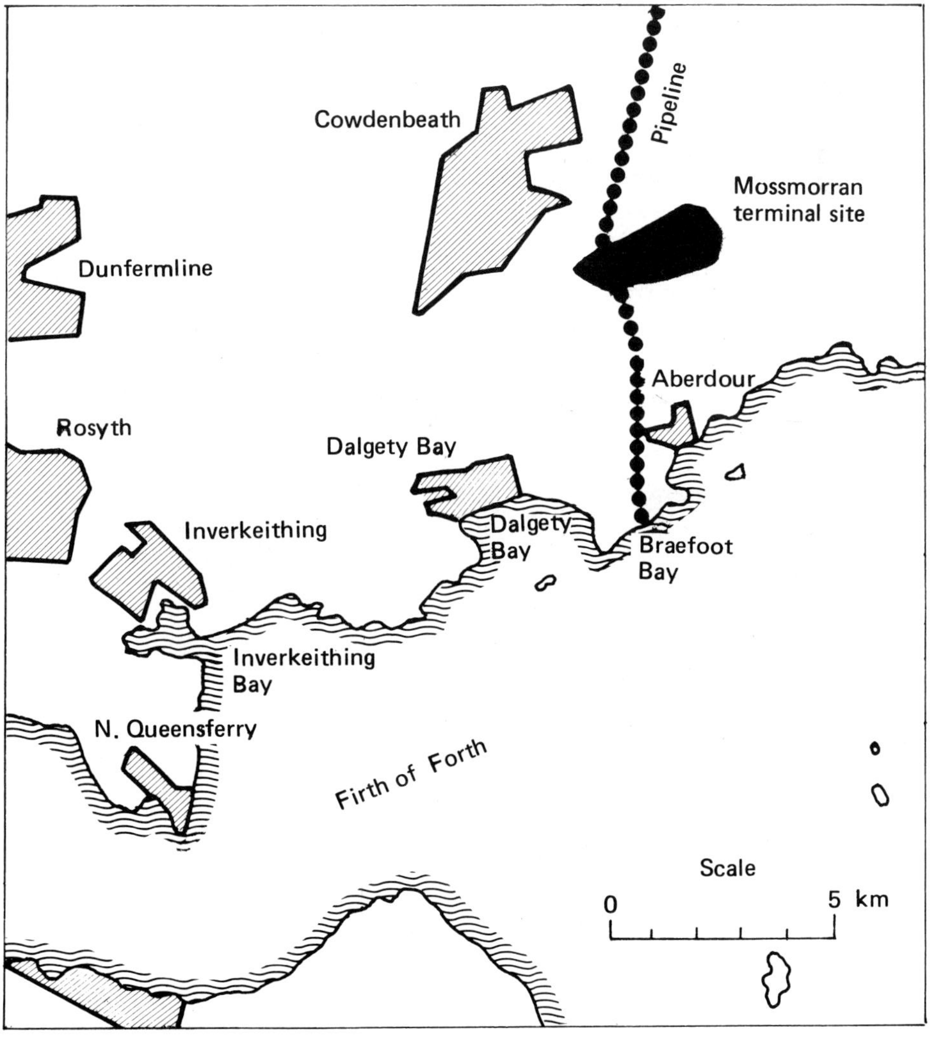

Figure 5.3. Mossmorran—Braefoot Bay.

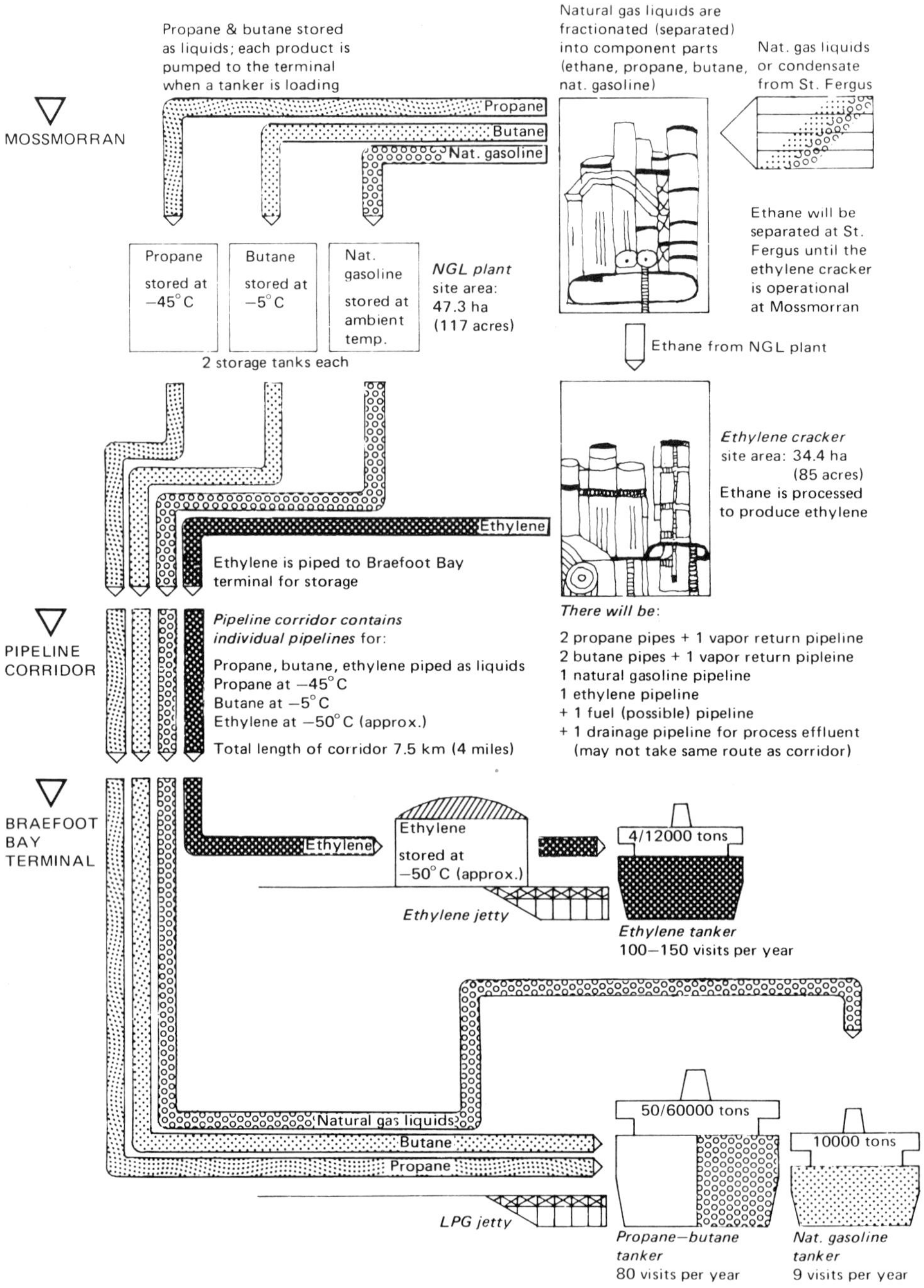

Figure 5.4. Interrelationships between the storage terminal and the shipping terminal for LEG (from Fife, Kirkcaldy, and Dunfermline District Councils 1977).

multinational corporation wanting to develop a 300-hectare site) requires planning permission. Official planning procedures in Scotland are set in train when a formal application for a particular development on a specific site is lodged. The onus of site selection rests firmly with the applicant.

Approval for planning applications may be given at the district, regional, or national government level, depending on the relationship of the project to the relevant government policies. The majority of applications are determined at the district level, and raise issues that are usually confined to the site in question. Those that raise issues of more than just district significance, specifically if they are contrary to or additional to the structure plan for the area (such as large industrial developments), may then be determined at the regional level. Finally, the Secretary of State may determine applications that raise issues of national significance, such as airports, power stations, or North Sea oil-related developments.

For large-scale projects such as Mossmorran—Braefoot Bay, the granting of planning permission involves a two-stage procedure, the first of which culminates in the granting or refusal of *outline* permission (i.e., approval in principle). It is this stage that, for the Mossmorran—Braefoot Bay development, corresponds to the decision process described in this case study. Outline planning permission is followed in a second stage (not reviewed here) by detailed consent. Once outline approval has been granted, it can be removed on very limited grounds of specified reserved matters; otherwise the government can face very substantial compensation claims. Thus outline permission, for most purposes, is regarded as full permission.

The level of detail required in the application for outline planning permission in this two-stage procedure has to be sufficient to establish the nature of the proposed activity so that the determining authority can establish in principle whether the site is suitable for accommodating the proposal. Design detail of the plant and specific support services that will be required (so-called reserved matters) are not usually investigated until the second stage, when outline planning permission for the overall development has been granted. In may cases this order appears to meet the interests of the developer and of the planning authority, since both parties may be reluctant to undertake detailed investigations before approval in principle to the overall development has been given. Thus outline planning permission relates only to the use of the land in principle, and not to any detail, even if this has been entered in support of the application.

Another feature to note about a development of the scale of the Mossmorran—Braefoot Bay facilities is that there are likely to be a number of informal consultations initiated by the developer with the district, regional, and national authorities in advance of the lodging of the official planning application. Such an exercise is directed initially at obtaining the basic information to enable the developer to decide whether the intended site is suitable. At the same time it enables the planning authority to assess the nature of additional information that it will itself require in order to be able to judge the acceptability of the application, e.g., whether any specific reports will need to be written or commissioned on particular aspects. Thus consultations undertaken during the pre-application period enable a more efficient processing of the formal application.

The major events and activities in the Mossmorran--Braefoot Bay decision process are illustrated in the PERT diagram in Figure 5.5, which divides the process into three rounds. The process as a whole took three years, from July 1976 to August 1979. There was no overall time limit on the duration of the decision, although certain procedures within it were bounded by statutory time constraints.

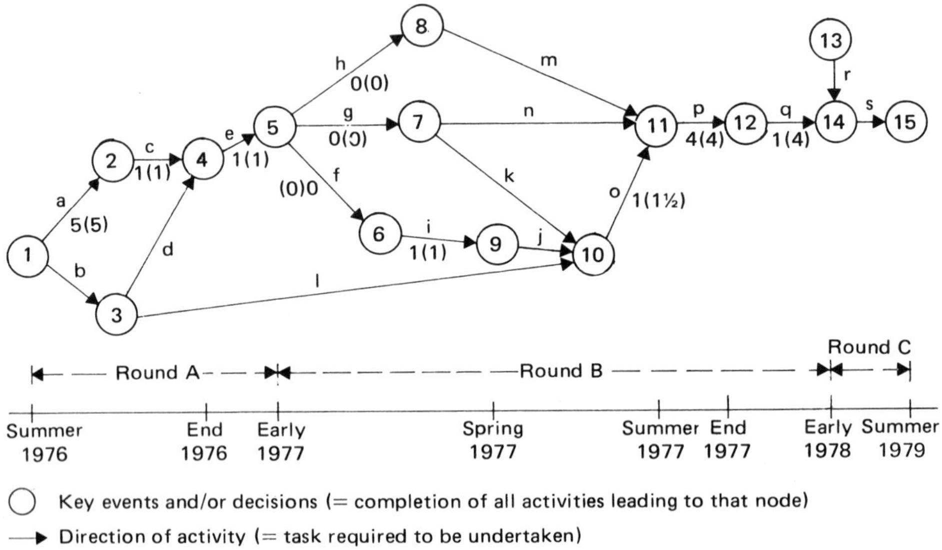

Figure 5.5. PERT diagram for the LEG decision process in the UK. Numbers along flow lines indicate the expected time in months for the relevant activities; those in parentheses indicate the actual times taken in this case.

Key events and/or decisions

① Oil companies show interest in Mossmorran—Braefoot Bay sites.
② Local authorities, Forth Ports Authority, and oil companies decide informally on suitability of site.
③ Local opposition articulated.
④ Public meetings held.
⑤ Formal planning applications lodged.
⑥ Regional/national levels notified.
⑦ Planning applications publicized.
⑧ Risk, environmental, and economic

(a) Informal consultations between oil companies and local authorities.
(b) Consternation amongst local residents about the project.
(c) Public meetings arranged by local authorities and oil companies.
(d) Local residents attend meetings.
(e) Oil companies prepare formal planning applications.
(f) Statutory notification set in train.
(g) Statutory publicity set in train.

impact analyses commissioned/prepared.
⑨ Applications called in by Secretary of State.
⑩ It is decided to hold a public inquiry.
⑪ Public inquiry is held.
⑫ Public inquiry report received by Secretary of State.
⑬ Radio sparks issue raised.
⑭ Secretary of State announces provisional approval but asks for further submissions on radio sparks.
⑮ Dialogue terminated; decision of approval announced.

(h) Consultations undertaken.
(i) Secretary of State reviews planning applications.
(j) Secretary of State awaits response to publicity.
(k) Signficant opposition articulated following publicity.
(l) Local residents further articulate opposition.
(m) Local authorities unanimously approve application and prepare for public inquiry based on formal analyses and consultations.
(n) Interested parties prepare cases for public inquiry.
(o) Statutory pre-inquiry activity coordinated by inquiry reporter.
(p) Inquiry report circulated for comment.
(q) Inquiry report considered by Secretary of State alongside "national interest" and other issues.
(r) Radio sparks issue brought to attention of Secretary of State.
(s) Interrupted dialogue between HSE and Action Group over radio sparks issue.

Round A. Round A began in July 1976, when the local authorities were approached by Shell and Esso. ① Informal contacts were established with other statutory authorities and interests concerning suitable sites for the facilities, ② as a result of which significant local opposition was articulated. ③ ④ Formal planning applications for the development of Mossmorran and Braefoot Bay were lodged with the district councils (Dunfermline and Kirkcaldy in this case, since Mossmorran straddles the district boundary), in January 1977. ⑤

Round B. This round began after the planning applications had been lodged, and initiated three interrelated lines of activity, as illustrated schematically in Figure 5.6.

(1) District councils must notify regional (Fife in this case) and national (Scottish Development Department in this case) authorities of applications whose impact may extend to the regional or national level, respectively. These higher authorities may, at their own discretion, "call in" the application, and thereby assume responsibility for determining the decision outcome. Applications that are not called in following such notification are passed back to the district level for determination. The Mossmorran—

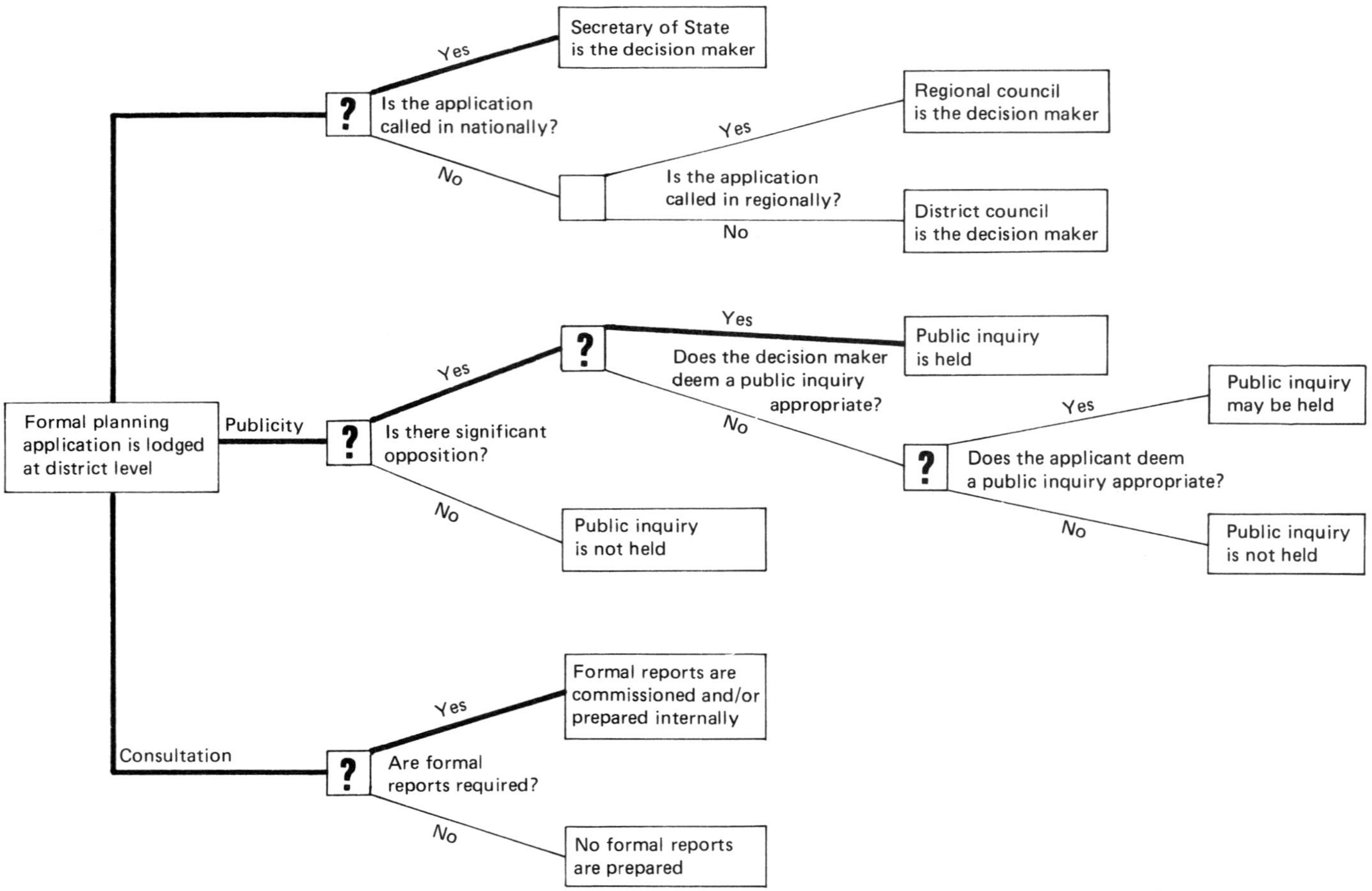

Figure 5.6. Possible statutory decision procedures in the UK following receipt of a planning application. Bold lines indicate

Braefoot Bay applications were called in at the national level in view of their importance to national energy policy, the national economy, and the environment of the Firth of Forth. (6) The ultimate decision taker was thus the Secretary of State for Scotland. Significant government interest at the regional level was also warranted and the three local authorities (regional and two district) coordinated their consideration of the applications, although each authority council ratified them individually.

(2) District councils publicize planning applications via notices, local press advertisements, and personal communications, so that any individuals or organizations (private or public), or other branches of government and public services who may be affected by them or have responsibilities that relate to particular aspects of the proposals, may make appropriate written representations before an official decision is made. In the case of the Mossmorran—Braefoot Bay applications there were many objections following such publicity, mainly on hazard and environmental impact grounds. (7) There was also some notable expression of support for the potential national interest and beneficial socioeconomic implications. Following informal consultations (July—December 1976), significant local opposition had already been articulated in advance of the formal publicity.

In the light of the response to the statutory publicity the Secretary of State may deem it suitable to hold a public inquiry into planning applications, affording all interested parties an opportunity to present arguments for or against the applications, and to cross-examine those of others. The proceedings may be similar to those of a formal court of law. Since the discretion to hold a public inquiry lies with the Secretary of State, it would have been legally possible, though politically difficult, to rule on the Mossmorran—Braefoot Bay applications without an inquiry. The Secretary does not witness the inquiry proceedings personally, and is not bound by any conclusions or recommendations that may stem from an inquiry. (10)

(3) Local authorities (district and regional) may seek professional advice from their own or from outside experts on the potential impacts of applications that may be significant from a planning point of view. In this case, Cremer and Warner, a private firm of chemical and engineering consultants, was commissioned to advise on hazard and environmental impact (Cremer and Warner 1977). (8) This step was taken to supplement the more general advice on potential hazard to be issued by the Health and Safety Executive, the statutory advisors and guardians of safety. The directors of planning of the three local authorities jointly prepared a report on socioeconomic impacts (Fife, Dunfermline, and Kirkcaldy District Councils 1977).

In addition to the requirement of planning permission for the facilities on land at Mossmorran and Braefoot Bay, a further statutory requirement was the granting of a jetty license by the Forth Ports Authority for the Braefoot Bay shipping terminal.

The public inquiry ⑪ was the centerpiece of round B of the decision process, which began with the lodging of formal planning applications. ⑤ In round A, considerations of the proposed developments had been essentially informal; ② ④ thereafter consideration became more formal, in line with statutory procedures, ⑥ ⑦ and the main formal analyses were prepared. ⑧ The public inquiry itself focused on local planning issues (as is quite usual); thus national government departments with a significant interest in the proposals (Energy, Industry, and Scottish Development) did not present their cases here, but issued brief written background statements and communicated their views to the Secretary of State through the normal course of official and unofficial (unobservable) interdepartmental contacts. In addition to their respective internal decisions and contacts, the other statutory authorities involved (Health and Safety Executive, local authorities, and the Forth Ports Authority) also presented their positions at the public inquiry. This became the main opportunity for acknowledged public debate of the issues at stake.

Round B ended when the Secretary of State received the public inquiry report and other relevant information. ⑫

Round C. Round C began after the public inquiry, when the issue of the potential hazard of the ignition of a vapor cloud from radio transmission break sparks came to light. ⑬ This was, most unusually, considered by the Secretary of State to merit further publicly acknowledged deliberation. Rather than reopening the inquiry as one possible course of action, public participation was invited in the form of written correspondence. ⑭ Thus, although the inquiry report had been received for consideration by the Secretary of State in November 1977, and an official decision had been expected shortly thereafter, consideration of the radio sparks issue postponed the final decision until August 1979. ⑮

PARTY CONCERNS

The large number of interested parties that were involved in the Mossmorran–Braefoot Bay decision process are listed in Table 5.1, and the potential impacts or concerns of these parties are given in Table 5.2. The notation and headings in this table are taken from the more comprehensive list of concerns presented in Table 2.1. The concerns were appraised differently by different parties, as indicated in the party/concern matrix in Table 5.3. Some concerns may be interpreted as relating directly to the utility or welfare of particular parties (denoted by ●), while others were of concern to parties representing areas in which they had explicit responsibility (denoted by ■). These concerns relate mainly, but not exclusively,

Table 5.1. Key parties in the LEG decision process in the UK.

Applicant	
Shell and Esso	Private international oil companies.
National government	
Secretary of State for Scotland	The decision taker, a national government cabinet minister, aided by legal and other advisors.
Scottish Development Department	Part of the Scottish Office acting on behalf of the Secretary of State, responsible for development control (an administrative function) and development planning (a policy function) in Scotland.
Departments of Energy and Industry	Government departments with significant policy interests in the developments.
Public inquiry reporter	A civil servant appointed by the Scottish Development Department to chair the public inquiry proceedings.
The Health and Safety Executive	Statutory UK guardian of safety, responsible for ensuring the safety of the public and employees potentially at risk from hazardous activities. The Executive's role is advisory in the planning stage, and enforcement in the operational phase.
Local government	
Forth Ports Authority	Statutorally responsible for leasing jetties and controlling traffic in the Firth of Forth, including responsibility for marine safety.
Local authorities	Fife region, Dunfermline and Kirkcaldy districts: host local authorities to the planned developments, democratically elected councillors advised by professional planners, and on hazard and environmental impact, by Cremer and Warner.
Others	
Aberdour and Dalgety Bay Joint Action Group	A local pressure group opposed to the developments, formed from the residents' associations of Aberdour and Dalgety Bay, two villages each with an estimated population of 4000, whose boundaries extend within one mile of the Braefoot Bay jetty. Articulate, predominantly middle class, suffering little unemployment, and considering that their views would not be represented other than through their own group, they were able to draw on indigenous legal and technical expertise, as well as advice from a professor of fire and safety engineering at Edinburgh University.
Conservation Society	A national environmentalist lobby, the Scottish branch of which made representations against the LEG plant.
The residents of Cowdenbeath	Cowdenbeath, the nearest town to the Mossmorran site, has a high rate of unemployment. Predominantly in favor of the plant on employment grounds, although they made little open articulation of this view within the formal decision process.

Table 5.1 (*continued*). Key parties in the LEG decision process in the UK.

Other parties	Other individuals and organizations were less prominent than those listed above, but not necessarily less concerned (see Macgill 1982). The main political parties supported the developments in principle (although a local MP raised questions in parliament on procedural aspects), and a few individuals from minority parties took an opposing view. The media leaned towards the objectors' arguments, although again there were exceptions.

Table 5.2. List of concerns relevant to the Mossmorran—Braefoot Bay decision

Concern		Interpretation of concern in the Mossmorran—Braefoot Bay context
National		
Economic benefits	N_4	Taxes, royalties, and balance of payment earnings
National costs	N_5	20% of capital construction costs of NGL plant and ethylene cracker
National image	N_6	The encouragement of oil-related developments
Energy policy	N_8	The efficient exploitation of the Brent oil and gas field
Regional		
Industrial development	R_2	The encouragement of industry and diversification of employment opportunities in Fife
Local		
Amenity and environment:		
land use	L_{11}	Potentially adverse impacts on these areas
marine pollution	L_{12}	
tourism, recreation	L_{13}	
local history	L_{14}	
Economic benefits:		
jobs, short-term	L_{22}	A maximum of 3500 construction jobs over three years
jobs, long-term	L_{23}	100, 250, and 1000+ jobs for each of the NGL plant, ethylene cracker, and downstream industries respectively
Population risk:		
shipping	L_{41}	
gas processing and storage	L_{42}	At Mossmorran
radio sparks	L_{44}	
jetty operations	L_{45}	At Braefoot Bay
Physical suitability of site	L_6	
Applicant		
Profit	A_1	

Table 5.3. Party/concern matrix.

Concerns		Parties: Applicant	National government					Local government		Other		
		Shell and Esso	Sec. of State	Scottish Devel. Dept	Depts of Energy & Industry	Inquiry reporter	HSE	Forth Ports Auth.	Local Auth- orities	Joint Action Group	Conser- vation Society	Cowden- beath residents
National			All concerns to be taken into consideration			All concerns to be taken into consideration						
Economic benefits	N_4				● ■				▲			
National costs	N_5	●								▲		
National image	N_6			● ■	● ■							
Energy policy	N_8	▲			● ■						▲	
Regional												
Industrial development	R_2	▲							●			
Local												
Amenity and environment:												
land use	L_{11}	■							●	●	▲	
fishing, marine pollution	L_{12}	■						■				
tourism, recreation	L_{13}							■	● ■	●		
local history	L_{14}								● ■	●	▲	
Economic benefits:												
jobs, short-term	L_{22}	●							● ■			●
jobs, long-term	L_{23}	●							● ■	▲		●
Risks to population:												
shipping	L_{41}	● ■					■	■	● ■	●	▲	
gas storage and processing	L_{42}	● ■					■		● ■	▲	▲	
radio sparks	L_{44}	● ■					■		● ■	●		
jetty operations	L_{45}	● ■					■		● ■	●	▲	
Physical suitability of site	L_6	●						■	● ■	▲	▲	
Applicant												
Profit	A_1	●										

● Concerns directly affecting party welfare.
■ Concerns for which party has direct responsibility.
▲ Concerns of strategic significance to party.

The presence or absence of any symbol in a particular cell is more important than the type of symbol. No weighting of the importance of particular concerns to particular parties has been attempted.

to the statutory (government) authorities. Some impacts, although not affecting the welfare or utility of individual parties, could be used as a basis for arguments by those parties for "strategic" reasons (denoted by ▲). For example, the Joint Action Group could use gas storage at Mossmorran, L_{42}, in this way; they were not likely to be directly affected by this impact, but it provided scope for argument against the plant and hence could augment their campaign.

From the static description given in Table 5.3, a dynamic representation of the decision process emerges by considering the interplay of party arguments over these concerns at various stages in the process. It turns out that the circles (●) identify some of the main causes for argument for or against the applications by parties primarily affected by them. It is to be noted, however, that not all such party concerns were openly debated by parties during the decision process — for example Shell and Esso over profit considerations, A_1, the population of Cowdenbeath over short-term jobs, L_{22}, and the Departments of Energy and Industry over energy policy, N_6. The squares (■) identify grounds for support or objections to the applications. In the case of potentially adverse impacts (amenity and environment, L_1, for instance), the squares represent arguments over whether the impact was or could be made acceptable, or be sufficiently balanced by other benefits. Not all such concerns were explicitly used as arguments during the decision process; for example, the Health and Safety Executive's concern over shipping, L_{41}. The triangles (▲) identify strategic arguments used to reinforce party positions.

THE MAMP FRAMEWORK

Following the MAMP framework, the UK decision process is divided into several rounds, each of which has a problem formulation; a starting point or initiating event; and interaction phase where party arguments are represented; and an outcome or conclusion. As noted in Chapter 2, rounds are loosely defined units, and there is no unique way to segment any given decision process. The Mossmorran—Braefoot Bay process is divided into three rounds for the purposes of this case study, as illustrated in Table 5.4 and the PERT diagram in Figure 5.5. Although this representation is designed to self-explanatory, some additional comment is appropriate.

Round A began when initial interest in the sites was expressed, and ended when formal planning applications were submitted. The end of round A was marked by a transfer of observable arguments from predominantly local to both local and national considerations, and by the entry of the ultimate decision taker, the Secretary of State for Scotland.

Only four parties appeared in round A. The residents of Cowdenbeath did not organize themselves as a participating party as such, either in this or in subsequent rounds. The provisional approval at this stage by the local authorities and by the Forth Ports Authority was important to the applicants.

Round B embraced the public inquiry — this was announced by the Secretary of State shortly after the "call in" — and ended with the Secretary of State's provisional decision of approval.

Round C was initiated by the raising of the radio sparks issue — a new concern that merited further open consideration. The reason why the sparks issue received so much prominence was procedural rather than substantive; it was not that the issue was a relatively serious one, but rather because it was only raised *after* the inquiry. The Action Group persisted with further arguments on other concerns during this round, but these were considered by the Secretary of State to be of a kind already dealt with at the inquiry, and therefore refused them acknowledged debate. Round C ended with the final decision of approval.

There had been strong support from both local and national government (Scottish Development Department, Departments of Energy and Industry, local authorities) to approve the applications, subject to satisfactory reassurance on safeguards against potentially adverse impacts, and this appears to have been little shaken by the relatively long delay over the radio sparks issue. A relatively extensive set of 48 planning conditions accompanied the approval of Shell's application, with an almost identical set for Esso's application. The most onerous of these from a safety viewpoint was that a full hazard and operability audit should be conducted before the facilities could be commissioned. This condition was strongly supported by many of the statutory authorities, who saw it as an integral step in ensuring the safety of the plant. The local authorities had suggested a similar condition at and before the public inquiry, following a recommendation by Cremer and Warner, that the audit should be to the council's satisfaction. The inquiry reporter, on the other hand, recommended that the audit should be to the satisfaction of the Health and Safety Executive. The Secretary of State decided that it must also be to *his own* satisfaction.

Not all arguments of the main parties are summarized in Table 5.4. There were undoubtedly a number of "unseen" interactions, both official and unofficial, and also vehement arguments from the Action Group on procedural aspects. The tabulations also mask the weight of concern by various parties about different impacts (see Macgill 1982). Moreover, only those arguments of the statutory authorities that relate to their approval in principle of the developments are reflected in Table 5.4, and not their more specific concerns over individual matters of detail. Individual parties did not substantially change their views on major concerns either during or between rounds.

The decision for the resolution of a large UK planning application would normally require two rounds, rather than the three indicated in

Table 5.4. The MAMP framework applied to the decision process on LEG in the UK.
(*a*) ROUND A: August 1976—early 1977.

I PROBLEM FORMULATION

Assumptions: It is both feasible and desirable to process gases from the Brent oil and gas field.

Question: Is Mossmorran—Braefoot Bay, in principle, a suitable site for a processing plant?

II INITIATION

Shell and Esso show interest in Mossmorran—Braefoot Bay site. (1) [a]

III INTERACTION

Party	Position	Arguments
Shell/Esso	In favor of site	Site designated for large-scale industry (R_2); environmental impacts can be made acceptable (L_{11}, L_{12}); availability of construction workforce advantageous (L_{22}); benefit of long-term jobs (L_{23}); *safety will be ensured* (L_{41}, L_{42}, L_{45}); site well suited physically to developments (L_6).
Forth Ports Authority	Site suitable	*Strict controls can ensure acceptable marine environmental impact and safety* (L_{12}, L_{13}, L_{41}, L_{45}); site well suited physically to developments (L_6).
Local authorities	In favor of applications; further advice sought on potential impacts; keen to encourage application (*c*) as well as (*a*) and (*b*)	Site designated for large-scale industry (R_2); potentially adverse, but on balance acceptable, environmental impacts (L_{11}, L_{13}, L_{14}); policy to encourage employment (L_{22}, L_{23}); *risks considered acceptable* (L_{41}, L_{45}); site physically well suited to developments (L_6).
Aberdour and Dalgety Bay Joint Action Group	Against applications	Significant environmental, historical, and amenity losses (L_{11}, L_{13}, L_{14}); *risks appear unacceptable* (L_{41}, L_{42}, L_{45}); Braefoot Bay site physically unsuitable (L_6).

IV CONCLUSIONS

Planning applications formally submitted.

Italics indicate arguments involving population risk.

[a] Circled numbers correspond with nodes on the PERT diagram in Figure 5.5.

Table 5.4 (*continued*). The MAMP framework applied to the decision process on LEG in the UK.

(*b*) ROUND B: January 1977—March 1978.

I PROBLEM FORMULATION

Assumptions: It is both feasible and desirable to process gases from the Brent field. Mossmorran—Braefoot Bay is potentially a suitable site. A public inquiry is to be held where concerns can be debated. Secretary of State for Scotland is the decision taker.

Question: Should the Mossmorran—Braefoot Bay site be officially approved?

II INITIATION

Planning applications are submitted, (5) processed, (6) (7) (8) and subjected to debate at the public inquiry. (11) Inquiry report subsequently submitted to Secretary of State. (12)

III INTERACTION

Party	Position	Arguments
Shell/Esso	In favor of site	Developments concur with national policy (N_8); site designated for industry (R_2); environmental impacts can be made acceptable (L_{11}, L_{12}); availability of workforce (L_{22}); benefit of long-term jobs (L_{23}); *safety will be ensured* (L_{41}, L_{42}, L_{45}); site well suited physically (L_6).
Departments of Energy and Industry	In favor of approval	Developments accord with energy policy (N_8); potential economic (N_4) and less tangible national benefits (N_6).
HSE	Site suitable	*Site suitable in principle for safe installations* (L_{42}, L_{45}).
Forth Ports Authority	Site suitable	*Strict controls can ensure acceptable marine environmental impact* ($L_{12}, L_{13}, L_{41}, L_{45}$); site well suited physically — no better site on estuary (L_6).
Local authorities	In favor of approval	Site designated for industry (R_2); potentially adverse, but acceptable, environmental impacts (L_{11}, L_{13}, L_{14}); jobs (L_{22}, L_{23}), plus multiplier effects of benefit; *accept advice that risks are acceptable*[a] (L_{41}, L_{42}, L_{45}).
Aberdour and Dalgety Bay Joint Action Group	Against site approval	Significant national costs (N_5); significant environmental and amenity losses (L_{11}, L_{13}, L_{14}); employees may be poached from existing industries (L_{23}); *risks unacceptable; little confidence in reassurances* (L_{41}, L_{42}, L_{45}); Braefoot Bay site physically unsuitable (L_6).
Conservation Society	Against site approval	North Sea resources should be conserved (N_8); significant environmental and amenity losses (L_{11}, L_{13}, L_{14}); *risks appear unacceptable* (L_{41}, L_{42}, L_{45}); site physically unsuitable (L_6).
Inquiry reporter	Recommends approval	National benefits must be accepted (N_3, N_4, N_6); beneficial local impact (R_2, L_{22}, L_{23}); acceptable environmental but small amenity losses ($L_{11}, L_{12}, L_{13}, L_{14}$); *weight of evidence suggests risk acceptable* (L_{41}, L_{42}, L_{45}); no better site on estuary (L_6).

IV CONCLUSION

The Secretary of State gives provisional approval to applications, subject to a wide range of planning conditions. Overriding national need outweighs potential environmental losses. No question of unacceptable risk emerging given Health and Safety at Work Act, and additional provisions, including those embodied in planning conditions. Sparks issue has been raised since the public inquiry; clarification of this issue required before final approval is granted.

[a] Denotes the submission of a formal risk study.

Table 5.4 (*continued*). The MAMP framework applied to the decision process on LEG in the UK.

(*c*) ROUND C: March 1978–August 1979.

I PROBLEM FORMULATION

Assumptions: The Secretary of State for Scotland has provisionally approved the Mossmorran–Braefoot Bay site.

Question: Do radio sparks pose an unacceptable risk?

II INITIATION

The Secretary of State considers that further investigation of the sparks issue is warranted.

III INTERACTION

Party	Position	Arguments
Health and Safety Executive	Site suitable subject to additional safeguards	*No reason to delay planning permission*[a] (L_{44}).
Local authorities	In favor of approval	*No reason to delay planning permission*[a] (L_{44}).
Aberdour and Dalgety Bay Joint Action Group	Against applications	*Sparks issue not resolved* (L_{44}); *significant additional evidence on other safety concerns*[a] (L_{41}, L_{42}, L_{45}).

IV CONCLUSION

August 1979: Secretary of State ratifies provisional approval indicated in Round B.

[a] Denotes the submission of a formal risk study.

Table 5.4, with no observable interaction following receipt of the inquiry report by the Secretary of State. One of the remarkable features of the Mossmorran–Braefoot Bay decision was that "new" evidence (the radio sparks issue) came to light *after* the public inquiry, and led to protracted consideration of the evidence in round C.

In rounds A and C, the public was able to participate through informal public meetings, extensive lobbying, and canvassing, as well as written representations, both to statutory authorities and to other interested individuals and organizations. Objections centered mainly on safety, the environment, and amenity concerns. In round B it was dominated by the formal public inquiry and the preparation of a convincing case of opposition by the Action Group's expert witnesses and other individuals and organizations who wished to express their views, and to question those of others. Safety emerged as their predominant concern. Notwithstanding their participation, the Action Group expressed a severe loss of confidence in the UK public inquiry system. That afterwards only those written representations dealing with the radio sparks issue were acknowledged as being relevant additional evidence, and that there were no substantial responses on other issues gave rise to their further disquiet and frustration.

Many members of the public who might have expected to be directly affected by the planned installations — notably the residents of Cowdenbeath — did not actively participate in the decision process for a number of reasons (many considered themselves adequately represented by the local authority councils, for instance). Those that did participate, however, articulated their views mainly through the Action Group, arguing on behalf of sections of the Aberdour and Dalgety Bay communities that were potentially most at risk.

The Aberdour and Dalgety Bay Joint Action Group was an extremely determined, articulate, and well organized pressure group that was self-funded and notable for the extent of its indigenous expertise — both legal and technical. The involvement of the Conservation Society was modest by comparison. The Action Group claimed that its technical expertise was superior to that of the statutory authorities, and called for more stringent safety requirements than those apparently considered adequate by those authorities. The Action Group was openly critical of the Cremer and Warner report, the competence of the Health and Safety Executive, and of the safety assurances given by the Forth Ports Authority. Much of the Group's concern over safety appears to have been intimately bound up with its concern over procedures, and it is these aspects that are of relevance in the present context. Fuller details of these aspects can be found in the main case study report (Macgill 1982), but in summary, the main concerns of the Action Group were:

- The decision procedures themselves, especially the site specificity of the process, the lack of opportunities for public participation, the lack of detailed information available, and the time and other resource constraints imposed on their own representations.
- The system of public participation in general, and the public inquiry system, particularly its unsuitability for handling decisions of technical complexity.
- The extent to which the safety of a proposed installation can (and should) be seen to be established during the decision process — e.g., in addition to (unseen) statutory safety scrutiny once official approval in principle has been granted.
- The problems of accountability that arise when the population considering itself most at risk is too small for normal democratic representation, and the single-issue nature of their concern is arguably unsuited to representation at general or local elections.
- The nature of the contribution of a public interest group, at its own expense, to societal risk management.

THE RISK ISSUE

The Mossmorran—Braefoot Bay decision process described in this chapter was geared to the granting of *outline* planning permission for gas processing, storage, and transhipment facilities. As such, the question at issue with regard to safety, framed by statutory procedures, was whether an acceptably safe installation of the type outlined could *in principle* be built on the chosen site. Risk studies were prepared by various parties in relation to this question, although at various stages the risk aspect appears to have become enmeshed with the more specific issue of whether the actual installations and associated activities would be acceptably safe.

The most important risk studies compiled were: a report by Cremer and Warner (1977); reports on radio transmission break sparks (see, for example, HSE 1978a); and the Action Group's shipping hazards report (1979). Other more modest hazard impact statements were prepared by individual parties as bases for their public inquiry evidence and, in the case of Shell and Esso, in support of their original applications, and in the case of Cremer and Warner, on radio sparks. Formal risk studies were *not* issued by the Health and Safety Executive (with the exception of those specifically on the radio sparks issue), since their public statements were required only to relate to whether an acceptably safe plant could be built in principle. The Executive's safety scrutiny would be made on the basis of the detailed plans of the facilities, but these would only become available after outline planning permission had been granted; only the outline plans were available during the decision process.

The three major risk studies were as follows.

(1) The Cremer and Warner report (1977) was commissioned by the local authorities towards the end of round A. It was written in response to a general brief asking for: (i) advice as to the acceptability of applications (*a*) and (*b*) in terms of hazard, including possible interaction effects and the layout of the proposed development; (ii) advice on the adequacy of information provided by the applicants; and (iii) possible recommendations for planning conditions. There was no statutory requirement that such a report should be prepared, but since hazard was a crucial element in the general cost—benefit balance of the applications at the local level, and because the advice from the Health and Safety Executive would be more general, the local authorities sought outside expert advice on this aspect. A small contribution to the cost of the report was contributed by the Scottish Development Department, who had an interest in this study.

The Cremer and Warner report was the most comprehensive risk study written during the decision process. It was necessarily restricted by the amount of design detail available at the outline planning stage: generic rather than specific design features had to be inferred at various stages. Consequence analysis calculations were used to estimate the probability of various types of failure (allowing for equipment malfunction and human error), estimates of which were expressed as low, very low, or extremely low. The overall conclusion reached by Cremer and Warner was that:

> There is no reason to doubt that the installations proposed for Mossmorran—Braefoot Bay cannot be designed, built and operated in such a manner as to be acceptable in terms of environmental impact and community safety — provided that reasonable and adequate safeguards are agreed and ensured.

Some specific recommendations regarding prudent layout and containment measures within the site were made in relation to applications (*a*) and (*b*), and for the evacuation of residents of a small housing estate at the edge of the development area. The report also suggested that a detailed technical safety audit should be completed on the plants prior to commissioning. Close scrutiny of the content of application (*c*) was also recommended.

The Cremer and Warner report was used by the local authorities to underpin their own judgment that the risk would be acceptable in light of the later enforcement role of the Health and Safety Executive and the good reputations of Shell and Esso. The report was assessed internally by the local authorities and by the Scottish Development Department and was made available for public inspection prior to the inquiry. The report was challenged by the Action Group with respect to its lack of detail, the qualitative criteria used, and notably the relatively scant attention paid to jetty operations at Braefoot Bay. The Action Group also questioned the impartiality of the consultants appointed by local authorities who appeared keen to support the application.

The public inquiry brought out a number of differences in the views of Cremer and Warner, the risk experts of Shell and Esso, the Health and Safety Executive, and the Action Group. With the exception of the Action Group, these were considered by the other parties to be differences related to matters of fine tuning rather than fundamental factors arguing against an acceptably safe plant. The Action Group, however, considered the differences to be more substantial, giving cause for the public to be fearful for its safety, particularly the hazard potential of open flammable cloud explosions (Rasbash and Drysdale 1977).

(2) The radio sparks issue (round C), although considered by all parties to be relatively less important than some other safety concerns, was given by far the most thorough treatment in terms of risk studies prepared during the decision process. The reports issued by the Health and Safety Executive were based on field experiments designed to assess the potential of radio transmission break sparks to ignite vapor clouds. These were undertaken in response to requests from the Action Group on this issue. The delay caused by the time taken to complete the experiments, and to circulate the results, was an important procedural feature of the Mossmorran—Braefoot Bay decision process.

(3) The Action Group's shipping hazards report, which was published during round C, was aimed to provide a quantitative assessment of risks to Forth communities from the proposed jetty operations at Braefoot Bay. The shipping risk element had received relatively little attention in the Cremer and Warner report, and none from the Health and Safety Executive at the public inquiry. The Action Group had long pressed for a quantitative assessment of risk, having suggested that a hazard probability of one in a million in any given year was a useful yardstick of risk acceptability, and noted that other parties implicitly accepted this yardstick, though without reference to an explicit quantitative assessment.

The shipping hazards report adopted methods used in an official assessment of the hazards at the Canvey Island gas and petrochemical complex (HSE 1978b), which had then just been published. It calculated that the annual probability of an incident at Braefoot Bay that could cause death or severe injury to members of the communities of Aberdour and Dalgety Bay was of the order of 10^{-3}. Lower probabilities were calculated for other Firth of Forth communities.

The report was considered by the Action Group to provide disquieting justification for their fears on the safety of the proposed installations. Although written by amateurs, its methods were verified by expert private consultants as being appropriate applications of those used in the Canvey Island report. Other parties inferred, however, that the use of Canvey "data" was inappropriate, but no written critique was made. The Action Group was concerned that there were no significant responses from statutory authorities on the substance of their shipping hazards report, or — in view of inferred criticisms — that a more sophisticated study of its kind was not undertaken before approval was granted. As an element of "post-inquiry" evidence, statutory procedures rendered this inappropriate. Thus the report did not stimulate open dialogue or debate with other parties in the decision process, although it did provide a focus for some third-party interest in the Action Group's campaign. Notwithstanding possible weaknesses in its technical content, this illustrates the point that the Group's concerns on safety were bound up with those over procedures.

CONCLUDING REMARKS

The risk studies used in the Mossmorran—Braefoot Bay decision process were part of a public debate on risk that was unbalanced and inconclusive. Given the philosophical interest in the problems being dealt with, this can hardly be unexpected. There are, however, a number of specific observations, aside from deeper philosophical considerations, that should be made. The debate was unbalanced in the sense that the emphasis given to various safety issues did not match the relative importance of those issues. For example, shipping hazards were given disproportionately little attention, whereas radio sparks were subject to extensive open scrutiny. The debate was inconclusive given that the interested parties could not reach agreement on criteria on which to judge safety. Also, many of the doubts about safety that were raised at the public inquiry, and in the quantity of post-inquiry evidence, remained unanswered.

Given that the debate was also severely limited by the amount of plant detail available, it is notable that some potentially important modifications in the handling of planning applications for major hazardous installations have occurred in the UK since the Mossmorran—Braefoot Bay decision. The outstanding arguments on safety were only partially resolved by requiring that a detailed hazard audit should be prepared to the satisfaction of the Health and Safety Executive and the Secretary of State prior to the commissioning of the installations, a planning condition that was unique at the time. This requirement was in addition to the nornal provisions of the Health and Safety at Work Act, which requires the Health and Safety Executive to scrutinize the general operation of the plant. This was a satisfactory resolution for the statutory authorities, but not for the Action Group, who called for greater stringency, a more explicit basis for the judgment of safety, and greater opportunity for open safety scrutiny during the decision process before they would be convinced of the safety of the proposed installations.

In retrospect, the Mossmorran—Braefoot Bay decision may be identified as something of a landmark in the handling of planning applications for major hazardous installations in the UK. The Health and Safety Executive are now more positively involved in the early planning stages — in defining the brief of advisory reports on safety and, if necessary, requesting more detail from the applicants about the proposed developments. Over and above any procedural changes that may have occurred, however, it remains true that the safety scrutiny of installations in the public eye during the decision process can only be a modest beginning of the detailed scrutiny (away from the public eye) they will receive from the Health and Safety Executive, in conjunction with industry self-regulation when the plants are under construction and in operation. A crucial consideration for some members of the public in reaching a judgment on risk acceptability may be their level of confidence in industry as self-regulators and the Health and Safety Executive as scrutineers. The procedural modifications that have occurred since the Mossmorran—Braefoot Bay decision are important not only in giving a more substantial role for the content of risk assessments — in terms of scope and depth of the safety issues covered — but also in providing a more positive demonstration of the effectiveness of the Health and Safety Executive as the guardians of public safety.

Notwithstanding these procedural changes, it is doubtful whether the Action Group would have been satisfied with the same final decision.

Although their fears on safety were bound up with criticisms of procedure, a number of outstanding concerns about safety (not shared by other parties involved) still remain. Indeed, a scenario could be constructed in which the recent procedural changes, allowing a more explicit basis for a debate on safety, could give rise to a more rather than less vigorous campaign. But that is another case study.

Chapter 6

The USA: Conflicts in California*

In the late 1960s, based on projections of decreasing existing natural gas supplies and increasing demand, several US gas companies began to seek additional supplies. In 1974, Western LNG Terminal Company, representing the terminal siting interests of three major utility companies, applied for approval of three LNG import sites on the California coast: Point Conception, on a remote and attractive part of the coast; Oxnard, a port city; and Los Angeles, a large harbor metropolis (see Figure 6.1). Western sought approval for all three sites in order to minimize the volume of tanker traffic at any one site, to separate ownership and control, and to reduce the risk of LNG supply interruption due to possible problems at any one location. The LNG would be shipped from Alaska's North Slope, Cook Inlet in southern Alaska, and Indonesia to the three sites. After nearly a decade of controversy, the utilities have announced that they will defer pursuing further their application for Point Conception, the one site remaining under active consideration, because California no longer needs to import natural gas.

SETTING THE STAGE

Energy planning in the US is decentralized in that initiatives for new energy facilities come in the form of applications to government agencies from energy supply and distribution companies. These companies are largely independent monopolies, but they are subject to regulation by federal, state, and regional agencies. Facility siting applications are also subject to approval from local and municipal planning and regulatory agencies. While federal agencies generally have authority for energy policy issues, those at the state and local levels usually have authority for land-use

*This chapter was written by John Lathrop and Joanne Linnerooth, based on a more comprehensive case study (Lathrop 1981, Linnerooth 1980).

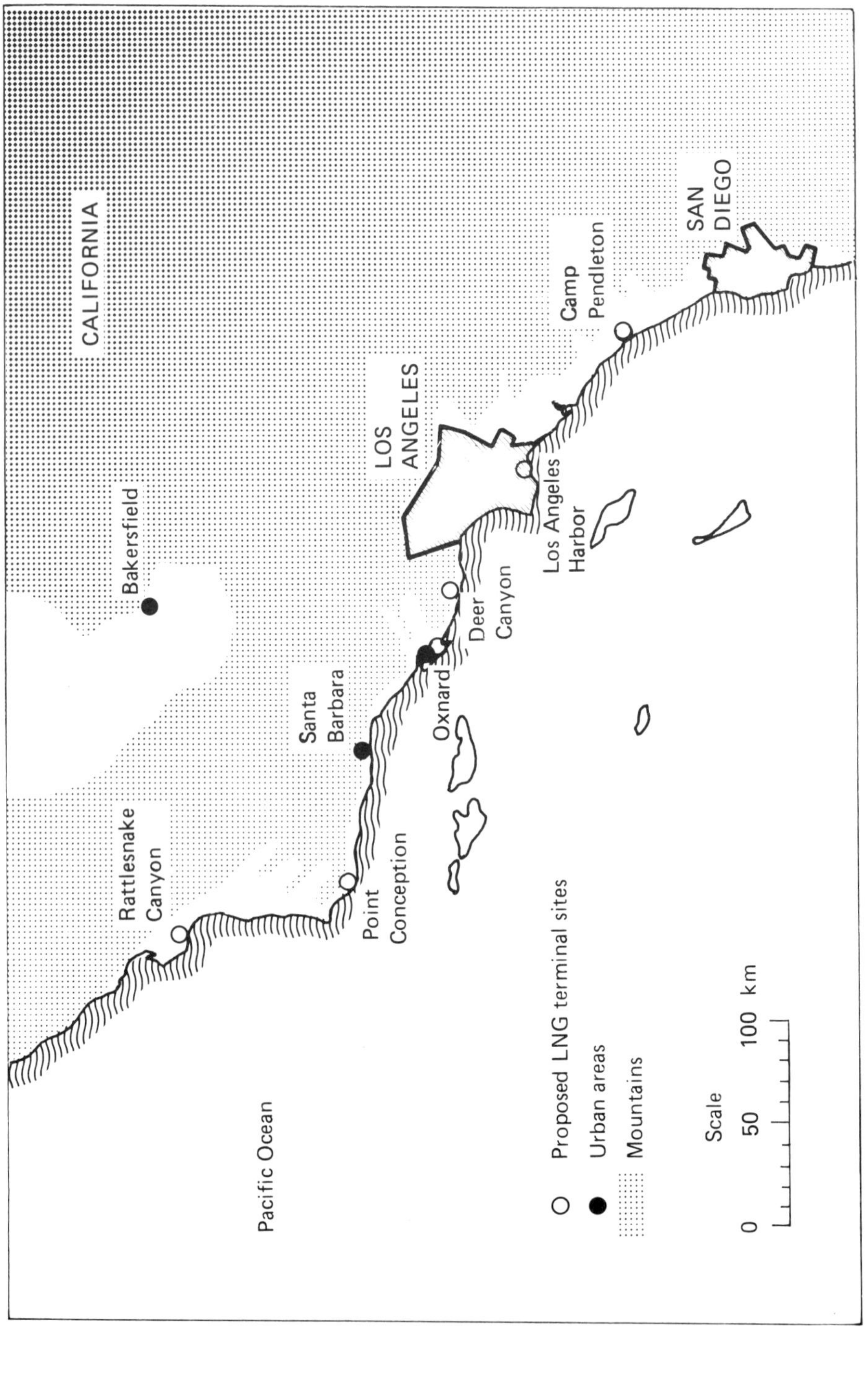

Figure 6.1. California, showing the location of the sites considered for the LNG terminal(s).

questions. Thus the division of authority for siting energy facilities is not clear. While no state has the right to impede national interest projects, it is rare for the federal government to impose its will over state protest.

At both the state and federal levels, quasi-judicial processes are usually used to arrive at important regulatory decisions. An administrative law judge presides over a set of hearings where representatives from interested parties (applicants, opposers, affected groups such as labor unions, etc.) present their cases. The staff of the government agency or commission (e.g., the California Public Utilities Commission or the Federal Energy Regulatory Commission) also presents its case. The judge then delivers his or her opinion to the person or group of commissioners heading the agency or commission, who make the final decision. They can agree with the judge, or disagree if they can establish defensible grounds for their case. The staff tends to have a technical and legal orientation, while the heads of the agency or commission are apt to be sensitive to political considerations.[1]

PERTINENT EVENTS

With a short introduction to the main actors, this section provides a brief overview of the sequence of events and decisions that comprised the decision process in California.

Table 6.1 lists the major parties, with a brief description of their roles in the decision process. *All* parties were essentially players in the game, differing chiefly in what levers each had available to manipulate, and in what interests and responsibilities each of them held. The formal interactions between the parties are shown in Figure 6.2. Perhaps the most notable informal interactions (including lobbying efforts and personal communications) included those between Western, public interest groups (such as the Sierra Club), and the California legislature. No local government bodies are included in this figure because the 1977 California LNG Terminal Siting Act precluded a central role for them.

The PERT diagram in Figure 6.3, as described in detail in Chapter 2, presents a flow chart of the key events and/or decisions that occurred during the California LNG siting debate. The process can be broken down into four separate, but by no means unique, rounds. These rounds are described below where the major decisions and events denoted by circled numbers correspond with the nodes in Figure 6.3.

Round A. In the late 1960s, the Pacific Lighting Corporation began to investigate gas supplies from Indonesia and Cook Inlet, Alaska. In 1972, a letter of intent was signed by PacIndonesia and Pertamina (the Indonesian state-owned oil and gas company) for the purchase of about 15 million m^3 of gas per day. After three years of price negotiations, the LNG contract was approved by the Indonesian government.

Table 6.1. Key parties involved in the California decision process.

Applicant	
Western LNG Terminal Company	A company owned by and representing the LNG terminal siting interests of two gas distribution companies: Pacific Lighting Corporation (southern California) and Pacific Gas and Electric (northern California). Until 1977 it also handled the California terminal portions of an LNG project for El Paso Natural Gas Company.
National government	
Federal Power Commission (FPC)	Essentially a financial regulatory agency with a mandate to regulate pricing policy and approve gas import projects. The FPC was superseded in 1977 by the Federal Energy Regulatory Agency (FERC) and the Economic Regulatory Administration (ERA) as part of the newly created Department of Energy.
Federal Energy Regulatory Commission (FERC)	Part of the Department of Energy (DOE), this commission was originally a typical economic regulatory agency, but under the National Environmental Protection Act (NEPA) its activities have expanded to generating Environmental Impact Statements (EIS) for each major application and certifying compliance with NEPA (best available alternatives, etc.). For LNG it must also certify compliance with the Natural Gas Policy Act (NGPA). In issuing Certificates of Public Convenience and Necessity, the FERC examines all aspects of a project, including costs, articles of incorporation, environmental impacts, and safety.
Economic Regulatory Administration (ERA)	While the FERC is largely technical in orientation, the ERA is the political regulator. As such, it concerns itself with the supply and pricing dependability of the source country, import price, balance of payments, and contractual structure of the marketing.
Office of Pipeline Safety Regulation (OPSR)	Sets technical safety regulations for the land portion of LNG terminals. It does not, however, consider each application as part of the permitting process.
Coast Guard (CG)	Sets technical safety standards for ships, ship operations, and the marine portion of LNG terminals. It rules on the acceptability of each application, and issues licenses for the LNG tankers.
State government	
California legislature	An elected body of representatives responsible for developing and approving laws pertinent to certain activities within the state.
California Public Utilities Commission (CPUC)	A typical US public utilities commission (PUC) with financial regulatory powers. As part of its duties it must determine whether costs incurred by a utility can be passed on to rate payers, and ensure adequate energy supplies to avoid related unemployment. Compared with other state agencies, such as the California Coastal and Energy Commissions (CCC and CEC), the CPUC tends to take

Table 6.1. *(continued)* Key parties involved in the California decision process.

	stands on capacity expansion closer to the utilities' point of view.
California Coastal Commission (CCC)	Charged with balancing the development and conservation interests in the management of the California coastline, the CCC has gained a reputation for deciding more often in favor of conservation than development, and tends to concentrate more than the CPUC on trade-offs between environmental quality and economics.
California Energy Commission (CEC)	The California Energy Resources Conservation and Development Commission was created in 1974 by both environmentalists and the utility interests for the promotion of conservation and alternative technologies. It was given the authority to issue power plant siting certificates as a way of streamlining siting procedures. The CEC serves as technical consultant to the CPUC on LNG matters. It also reports to the legislature on future state energy needs and supplies.
Local authorities	
Los Angeles City Council	An elected body with responsibility for a variety of local activities such as industrial development, employment, housing, health and safety, and environmental protection. LA is a large city with many diverse interests competing for support.
Oxnard City Council	An elected body with responsibility for a variety of local activities such as industrial development, housing, health and safety, and environmental protection. Compared with LA, Oxnard is quite small, with a population of 100 000. Oxnard does not have any major sources of income for its tax base.
Santa Barbara County	Point Conception is located in the county of Santa Barbara. It has rigorous environmental quality standards and a long, challenging procedure for industrial facility approval.
Other interested parties	
Sierra Club	A national environmentalist organization with numerous state and local chapters. It represents a large membership of citizens concerned with the protection and preservation of the environment. The Club often forms coalitions with other environmentalist groups to advocate more effectively its concerns.
Hollister and Bixby Ranch Associations	Two relatively wealthy groups of landowners adjacent to the Point Conception site.

Meanwhile, Western LNG Terminal Company had been formed as a subsidiary of Pacific Lighting Corporation and Pacific Gas and Electric for the purpose of planning and building two import terminals. After somewhat limited site screening, the Port of Los Angeles was chosen to receive gas from Cook Inlet, and Oxnard was chosen to receive gas from

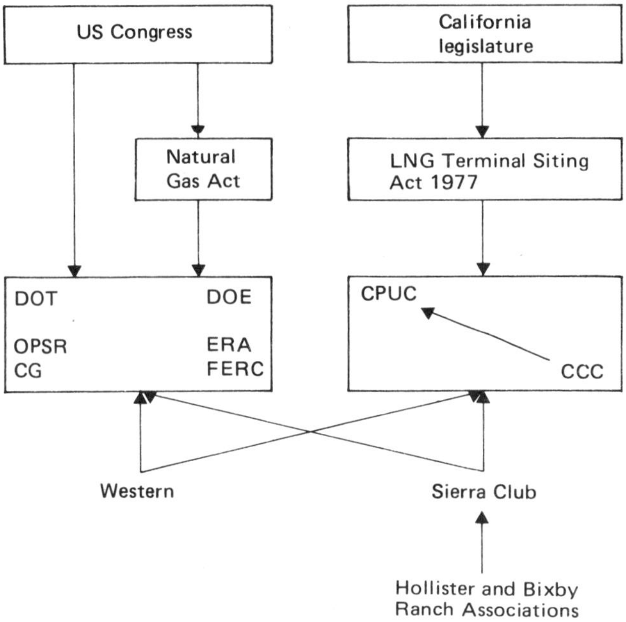

Figure 6.2. Interactions between the main interested parties in the California LNG decision process.

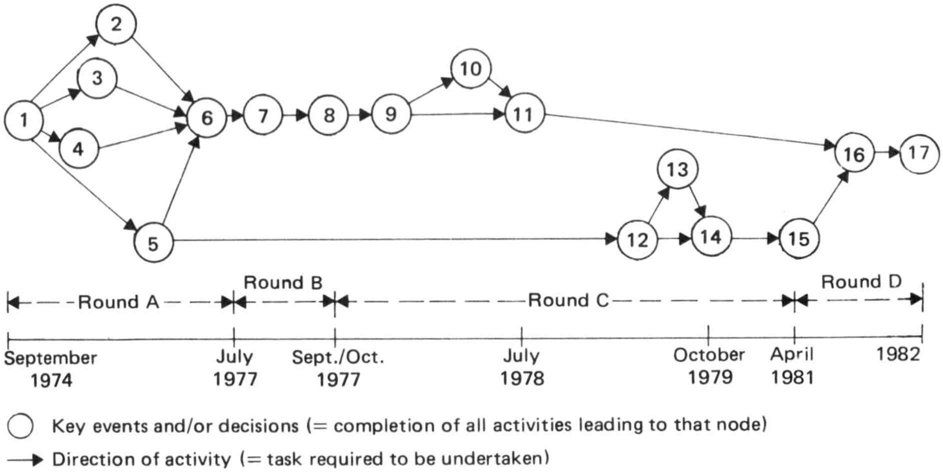

Fig 6.3. PERT diagram for the LNG decision process in the US.

Key events and/or decisions

(1) Western files applications with the Federal Power Commission (FPC) for approval of three sites: Point Conception, Oxnard, and Los Angeles.

(2) Oxnard City Council's response to an LNG terminal at Oxnard is uncertain, influenced by the SES risk assessment, which expresses concern over the potential risks.

③ LA City Council is in favor of an LNG terminal at LA, stressing the economic benefits and maintaining the acceptability of the risks.
④ The California Coastal Commission (CCC) suggests that Western actively pursue at least one remote site and one nonremote site (1977).
⑤ The FPC staff opposes LA due to an earthquake fault in the vicinity. The FPC risk assessment favors both Oxnard and Point Conception, but the US President's National Energy Plan calls for remote siting (summer 1977).
⑥ Western perceives a stalemate — the key parties at all three levels of government favor different sites — and fears that none of the sites will be approved.
⑦ Western, the utilities, other businesses, and labor unions press for legislation to expedite the decision process (July 1977).
⑧ State legislature passes the LNG Terminal Siting Act of 1977, which requires remote siting, and provides for "one-stop" licensing, whereby the CCC ranks sites and California Public Utilities Commission (CPUC) makes the final selection (September 1977).
⑨ Western submits application for an LNG terminal at Point Conception under the Siting Act (October 1977).
⑩ The CCC ranks Point Conception third out of four alternatives that meet the remote siting criterion.
⑪ The CPUC approves Point Conception on condition that additional seismic data are favorable (July 1978).
⑫ The federal government passes the Natural Gas Policy Act (NPGA) in 1978 to rectify the disparities created by previous regulations between unregulated intrastate and regulated interstate markets. The FERC is given primary responsibility for administering and enforcing the NGPA, which outlines some of the criteria by which both FERC and the ERA must make their decisions.
⑬ The ERA approves LNG imports, recognizing the need for natural gas and determining the source country supply interruption risk to be acceptably low (conditional approval December 1977; final approval September 1979).
⑭ The FERC approves the Alaska part of the project. The FERC staff prefers Oxnard, but the Commissioners approve Point Conception to avoid confrontation with the state legislature, which had already eliminated Oxnard (October 1979).
⑮ The Washington DC Court of Appeals decides in favor of a petition requesting that the FERC approval of Point Conception be reconsidered in the light of additional seismic risk data (April 1981).
⑯ The FERC and the CPUC hold concurrent hearings on seismic risk (1980 to September 1982).
⑰ Pacific Gas and Electric and Pacific Lighting Corporation announce that they will defer construction of the terminal until an unspecified date, since California no longer needs imported natural gas.

Indonesia. In 1972, Western extended its role to include building and operating an LNG terminal for the El Paso Natural Gas Company, which was planning a project to receive gas from Alaska's North Slope. Because El Paso had a corporate policy of not siting an LNG facility within ten miles of a populated area, the remote Point Conception (Little Cojo Bay) site was designated to receive gas from the North Slope.

In 1974, applications for these three sites were filed with the Federal Power Commission (FPC),[2] ① marking the beginning of round A, since the process had formally entered the political arena. In support of the densely populated Los Angeles and Oxnard sites, Western commissioned two risk assessment studies (SAI 1975a, b), both of which found the safety risks to be extremely low. These reports generated estimates of the probabilities that members of the public would be killed during any one year from terminal operations, based upon probabilities of marine and onshore LNG accidents and bad weather conditions. In the usual manner, these risks were compared with other possible causes of death (e.g., ill health and occupational hazards), and the reports concluded that the risks were extremely low. As required by federal law, environmental impact statements for the sites were prepared by the FPC. Oxnard City Council also commissioned a separate study of the environmental effects of LNG (SES 1976). In addition to probabilistic measures, this study expressed the risks of the project in terms of worst-case scenarios showing up to 70 000 casualties in the event of an accident, and this "electrified" opposition to the terminal (Ahern 1980).

Because of the absence of federal regulations relevant to LNG terminals, the difficult task of ensuring the safe operation of such a facility fell on the shoulders of the inexpert local authorities. Consequently, much of the blame for the uncertainties and problems surrounding LNG terminal siting has been seen to lie with the federal agencies (Ahern 1978).

Although during round A federal and local approval of a site was viewed as necessary,[3] the final approval was vested in a state agency — the California Coastal Commission (CCC). Created in 1976, the CCC is composed of 12 lay people from a variety of backgrounds serving part time, with responsibility for the protection of the California coastline. After much deliberation, the CCC decided that the risks of an LNG terminal were so uncertain that it could only approve a remote site at that time. However, when and if those risks were found to be acceptable, it could then approve a nonremote site. Consequently, in March 1977, the CCC advised Western to pursue at least one remote site and one nonremote site.

At the federal level, the FPC (which later became the Federal Energy Regulatory Commission, FERC) was in favor of the Oxnard site, but the US President's National Energy Plan called for the remote siting of LNG terminals. The FPC (FERC) was also deemed likely to deny the Port of Los Angeles site because of the existence of an earthquake fault, even though this site was favored by the local authorities. Again, at the local level, the Oxnard authorities seemed increasingly unlikely to approve a terminal. ② Western then faced a complex and lengthy approval process with Santa Barbara County, which held approval authority over the Point Conception site.

At the state level, it seemed unlikely that the CCC, placing priority on public safety, could be convinced that an LNG terminal was safe enough for either Oxnard or Los Angeles. But the CCC also faced problems in approving the remote Point Conception site, where the marine life, kelp beds, surfing breaks, and spectacular views represented the types of resources the Commission had been created to protect. ④ In addition, this site was being actively opposed by the Bixby and Hollister Ranch Associations who owned the adjacent land. To complicate an already complex situation, the Sierra Club opposed LNG on several fronts. They argued that California did not need the gas, but if it were imported it should go to Oxnard, because Los Angeles seemed unsafe, and the facility should be kept off the beautiful Point Conception site. However, the Club changed its stand in 1977 to oppose the Oxnard site on grounds of population risk, and so opposed all sites.

In summary, Western faced the possibility of failing to obtain the needed approvals for any of the three sites because the key parties at all levels of government favored different sites. ⑥

Round B. In view of this impending stalemate the utility companies turned to the state legislature for help. Their goal was to remove approval authority from the many local interests and the CCC, and to place it in the hands of the more congenial California Public Utilities Commission (CPUC). This agency is the principal state body involved in power plant issues, primarily in setting electricity rates.

The legislation being considered at that time (California Assembly Bill AB220), introduced by Assemblyman Goggins, was not acceptable to the utility companies. Although it would have given the CPUC exclusive authority to approve a proposed LNG facility, the CPUC was forced to consider the feasibility of both remote land-based and offshore sites. In addition, the bill required that the CCC and the California Energy Commission (CEC) offer second opinions on the feasibility decision. The CEC was known to oppose the CPUC on the question of LNG for California, and in its 1977 policy report to the legislature, the CEC raised questions about LNG safety, need, and costs. In the opinion of Western, this bill would have effectively prevented the siting of LNG facilities in California (Western 1978), so Western, together with other business and labor interests, went to battle for a rival bill (SB1081), which vested the CPUC with sole authority for approval, i.e., with one-stop licensing authority, precluding any real interference from the CEC. ⑦

The resulting legislation, known as the LNG Terminal Siting Act of 1977, was a compromise between the environmentalists, who supported consideration of offshore sites, and those who saw an urgent need for an LNG facility to ensure energy and jobs. The CPUC was chosen over the more conservation-minded CEC as the agency with state licensing authority, pre-empting local governments. As a bow to the conservationists, the CCC was given the mandate to choose and to rank possible sites, and to pass these rankings on to the CPUC. ⑧ It was agreed that the site would not be offshore, as some environmentalists wished; nor could it be in a populated area, as the gas utilities wished. Indeed, the required population density was strictly defined: there should be no more than an average of 10 people per square mile within one mile of the terminal, and no more than 60 people per square mile within four miles of the terminal.

Round C. In accordance with the LNG Terminal Siting Act of 1977, Western submitted an application to build a terminal at the remote Point Conception. ⑨ The CCC then evaluated 82 sites, 18 of which were nominated by the public. The CCC was required by law to rank the sites proposed by the applicant. Of these 82 sites, only four, including the third-ranked Point Conception, met the population density standards,[4] and were not found to be infeasible on the basis of adverse wind and wave conditions, earthquake faults, soil conditions, or other factors. The CCC ranked these sites and passed its evaluation on to the CPUC, which eliminated all but Point Conception, concluding that transients (campers, etc.) near the sites, on roads, and at public parks, made the other sites unsafe, and that the approval for these sites would involve unacceptable delay.

This, however, was not the end of the story. During the course of the screening process, the opposition (namely the Hollister and Bixby Ranches) presented evidence of earthquake faults at Point Conception. The CPUC only conditionally approved the site, stating in its July 1978 decision (the deadline date set by the 1977 Act) that this approval was conditional on Western's demonstration that the faults presented an acceptable risk to the terminal. ⑪ The CPUC also called for further wind and wave studies, but these studies did not generate any controversial data.

Concurrent with the state proceedings, Western had filed with the federal government for a license to import gas to Point Conception. The federal government was now operating under the conditions of the Natural Gas Policy Act (NGPA) of 1978, which was intended to set prices for both unregulated instrastate and regulated interstate markets, to remove outmoded regulatory burdens on the sales of interstate markets, and to provide incentives for risky exploratory drilling, while also restraining price increases on previously discovered gas. ⑫ The Federal Energy Regulatory Commission (FERC) was given primary responsibility for administering and enforcing this legislation, but approval of imports remained with the Economic Regulatory Administration (ERA). Both departments were in the newly created Department of Energy, so the PacIndonesian file was transferred from the ERA, which approved the importation of LNG to California, ⑬ to the FERC, which undertook an extensive environmental impact assessment. Although the staff of the FERC preferred Oxnard, the Commission decided in favor of Point Conception to avoid further confrontation with the California legislature. ⑭

Round D. On the basis of the need to consider additional seismic risk data, and continuing questions about California's need for LNG, opponents of LNG appealed against the FERC decision at the federal level. The

Washington DC Court of Appeals remanded the case back to the FERC, requesting a decision after reconsidering the case in light of additional seismic risk data. [15] The FERC and CPUC held concurrent hearings on seismic risk at Point Conception, and the CPUC ruled that the risks of the site were sufficiently low to permit construction. [16] However, it appeared likely that the CPUC would require additional hearings on the question of whether California still needed an LNG terminal. But these hearings will not now be necessary since the utilities have announced their decision *not* to build the terminal in the near future, but to keep open the option of constructing a terminal at Point Conception if the need should arise — perhaps in the 1990s. [17]

PARTY CONCERNS

To structure the siting process we need to have a good understanding of the different concerns of the interested parties. For the LNG problem, three categories of concern are relevant: risk aspects, economic aspects, and environmental aspects. Table 6.2 depicts a party/concern matrix showing the main concerns of the parties involved over this eight-year period.

The concerns listed have been selected to reflect the nature of the debate, that is, to reflect those concerns expressed by the parties involved, rather than to characterize the concerns in an analytical manner. For example, population risk involves the risk to life and limb borne by neighbors of the LNG terminal due to accidents, including those induced by earthquakes. Earthquake risk, which involves both population risk and supply interruption risk due to earthquakes, is included as a separate attribute since it was handled as such in the process. Supply interruption risk due to need refers to shortfalls in the supply of natural gas due simply to demand exceeding supply.

The crosses in Table 6.2 indicate the primary concerns of the interested parties. Naturally, many of the parties cared about aspects of the siting decision that are not marked as primary concerns. While "hidden concerns" may have motivated parties, they did not surface as explicit arguments and so these are not identified as primary concerns in this table.

For example, the FERC, the state legislature, and the CPUC were interested in consumer concerns, environmental quality, and risks. The ERA, on the other hand, was restricted by its mandate to a concern for price and particular types of supply interruption risk. A technical regulator, such as the Office of Pipeline Safety Regulation (OPSR), was limited to considering particular design features, while the CCC was specifically charged with preserving coastal resources, although it also displayed a concern for population risk. Local authorities and Western were naturally primarily concerned with those factors that directly affected

Table 6.2. Party/concern matrix.

		Parties									
		Applicant	Government							Other	
			Federal			State			Local		
Concerns		Western	FERC	ERA	OPSR	Legislature	CPUC	CCC	Municipal	Sierra Club	Neighbors
Regional											
Air quality	R_1		x			x	x			x	
Local											
Amenity and environment:											
land use	L_{11}		x			x	x	x	x	x	x
fishing, marine	L_{12}		x				x	x	x	x	x
tourism, recreation	L_{13}		x				x	x	x	x	x
Economic benefits:											
taxes, business revenue	L_{21}								x		
jobs, short-term	L_{22}					x			x		
jobs, long-term	L_{23}					x			x		
prestige	L_{26}								x		
Risk to population:	L_4		x		x	x	x	x	x	x	x
earthquakes	L_{43}		x		x	x					x
Risk to workers	L_5		x		x		x				x
Consumer											
Price of gas	C_2	x	x	x		x	x				
Supply interruption risk:											
due to need for gas	C_{31}	x	x	x		x	x				
due to source country	C_{32}	x		x		x	x				
due to weather	C_{33}	x	x			x	x				
due to accident	C_{34}	x	x			x	x				
due to earthquakes	C_{35}	x	x			x	x				
Applicant											
Profit	A_1	x									
Control over sources	A_2	x									
Image	A_3	x									

FERC: Federal Energy Regulatory Commission
ERA: Economic Regulatory Administration
OPSR: Office of Pipeline Safety Regulation
CPUC: California Public Utilities Commission
CCC: California Coastal Commission

them or their constituencies, although Western was also concerned with supply interruption risk. While the Sierra Club was interested in environmental quality and population risk, neighbors to the site concentrated more on local environmental quality and the risk to themselves both as residents and workers.

THE MAMP FRAMEWORK

The siting decision process in California, which was characterized by four rounds, will be discussed in more detail in accordance with the MAMP framework. The main elements of these rounds are listed in Table 6.3.

The application raised two central questions, and these framed the problem addressed in round A: Does California need LNG? and, if so, which (if any) of the proposed sites is appropriate?

The agenda for discussion was narrowly defined at this stage. The wheels of the process were set in motion, not by a broad-based energy-policy question initiated in Washington, but by a proposal from industry for three preselected sites. The importance of this process where the initiative was taken *first* by industry in preselecting the agenda for debate cannot be overemphasized. The initiating proposal formulated the problem as "Should the proposed LNG sites be approved?", and *not* "Should California have an LNG terminal in view of the alternatives, costs, risks, etc.?" Setting the agenda in this manner did not preclude the "need" question from entering the debate, but it did ensure that the question was only considered in the context of a siting application.

Table 6.3(*a*) also indentifies the relevant interested parties involved in the interaction phase of round A; those parties with formal decision responsibility are marked with a cross (+). The primary concerns that entered the ensuing debate included the need for LNG, the risk of supply interruption, land use, and the health and safety risks to the local population. The need for LNG or the risk of an interruption in the supply of natural gas (C_{31}—C_{35}) supported the location of a terminal at at least one of the three proposed sites. Environmental and land-use considerations (L_{11}) suggested a nonremote site (Los Angeles or Oxnard), while the risks to the population (L_4) argued for siting the terminal in a remote area (Point Conception). Finally, concerns about earthquake risk (L_{43}) resulted in opposition to the Los Angeles site, which had been found to be crossed by a significant fault.

The interaction phase of round A detailed in Table 6.3 indicates the concerns used as *arguments* by each of the major parties. It is important to distinguish these concerns from those listed in Table 6.2, which specifies those attributes that were of *primary* concern to each party. For example, although Western was concerned with both profit considerations and supply interruption risk, its arguments in support of each site stressed *only* supply interruption risk.

Table 6.3. The MAMP framework applied to the decision process on LNG in the US.

(*a*) ROUND A: September 1974—July 1977.

I PROBLEM FORMULATION

Assumptions:	It is technically feasible to import LNG from Indonesia and Alaska into California.
Questions:	(1) Does California need LNG? (2) If so, which of the proposed sites is appropriate?
Process constraints:	There is a standard siting procedure developed from experience with smaller-scale and/or less novel facilities.
Outcome constraints:	A contract has been signed for natural gas, so timely siting is desirable.

II INITIATION

Western files with the FPC for approval of three sites: Point Conception (PC); Oxnard (OX), and Los Angeles (LA). ① [a]

III INTERACTION

Party	Position	Arguments
Western	For PC, OX, and LA	Minimize supply interruption risk (C_{31}); *low population risk** (L_4).
FPC (FERC)$^+$	For PC and OX, against LA	Minimize supply interruption risk (C_{31}); *high population risk due to earthquake fault at LA** (L_{43}).
CEC	Against PC, OX, and LA	Low supply interruption risk (C_{31}).
CCC$^+$	For pursuing both remote and nonremote sites	*High population risk* (L_4); environmental quality, land use (L_{11}).
LA City Council$^+$	For LA	Taxes (L_{21}); *low population risk* (L_4).
OX City Council	For OX (?)	Taxes (L_{21}); *low population risk* (L_4) (?).
Santa Barbara County	Against PC (?)	Environmental quality (L_{11}).
PC neighbors	Against PC	Environmental quality (L_{11}).
OX citizens	Against OX	*High population risk** (L_4).
LA neighbors	Against LA	*High population risk** (L_4).
Sierra Club	Against PC and LA, for OX[b]	Environmental quality (L_{11}); air quality, (R_1); *low population risk at OX* (L_4); *high population risk at LA* (L_4).

IV CONCLUSIONS

Key decisions

(1) *CCC* advises Western to pursue PC. ④
Procedure: internal among staff, 12 lay commissioners.
Key tradeoff: population risk outweighs environmental quality (land use).

(2) *FPC* staff oppose LA on seismic risk ground. ⑤
Procedure: position developed by staff, never brought to hearing.
Key trade-off: seismic risk to population outweighs supply interruption risk.

Applicant perceives a stalemate, i.e., that no site is approvable without long delays. ⑥

\+ denotes decision power.
* denotes supported with technical risk analysis.
(?) denotes uncertain position.
Italics indicate arguments involving population risk.
[a] Numbers in circles refer to nodes in the PERT diagram in Figure 6.3; this is applicable to all rounds in this table.
[b] In 1977 the Sierra Club shifted its position and opposed Oxnard.

Table 6.3. (*continued*) The MAMP framework applied to the decision process on LNG in the US.

(*b*) ROUND B: July 1977—September/October 1977.

I PROBLEM FORMULATION

Assumption: It is difficult, if not impossible, for Western to gain approval for a site under existing procedures.

Questions: (1) How should need for LNG be determined?
(2) If need is established, what changes are required in the process in order to site an LNG facility?

II INITIATION

Pressure is brought to bear on state legislature by Western, utilities, business, and labor unions to pass legislation implementing a siting procedure that is more amenable to siting an LNG terminal. ⑦

III INTERACTION

Party	Position	Arguments
Senator Alquist	For OX	Minimize supply interruption risk (C_{31}).
CPUC, Western, business, labor	For OX	Minimize supply interruption risk due to need (C_{31}).
Assemblyman Goggins	Against OX, for an offshore site	*High population risk** (L_4).
CCC	Against OX, for an offshore site	*High population risk* (L_4); environmental quality, land use (L_{11}).

IV CONCLUSIONS

Key decisions

(3) State legislature passes LNG Terminal Siting Act (SB1081). ⑧
Procedure: Closed meetings to set principles, draft (Sierra Club excluded). Floor vote.

The Act states that California needs an LNG terminal, and calls for an accelerated siting procedure in which the CCC will rank sites and the CPUC will select from this ranking (one-stop licensing). In addition, the site must not be offshore but must be remote.

Act features	*Corresponding trade-offs*
Accelerate siting	Economic health over environmental quality, local sovereignty.
One-stop licensing	Coherent policy, supply interruption risk over local sovereignty.
One-stop is CPUC	Supply interruption risk over environmental quality.
No offshore site	Supply interruption risk over environmental quality.
Site must be remote	Population risk over environmental quality.

Table 6.3. (*continued*) The MAMP framework applied to the decision process on LNG in the US.

(*c*) ROUND C: September/October 1977—April 1981.

I PROBLEM FORMULATION

Assumptions:	(1) California needs an LNG terminal. (2) A site must be selected according to procedures specified by SB1081.
Question:	Which site is appropriate?
Process constraints:	As specified by SB1081, the CPUC must select a site from those ranked by the CCC, subject to some conditions, by 31 July 1978.
Outcome constraints:	The site must be remote and land-based.

II INITIATION

Applicant files for approval of Point Conception site (applications for both PC and OX remain technically "alive" at federal level). ⑨

III INTERACTION

Party	Position	Arguments
Western	For PC	Minimize supply interruption risk (C_{31}); no seismic risk (C_{34}).
ERA	Approves import to PC	Minimize supply interruption risk (C_{31}); improves air quality (R_1); no seismic risk (L_{43}).
FERC+	For OX	Does not degrade environmental quality (L_{11}); *low population risk at OX* (L_4).
	For PC	Avoid state—federal confrontation.
CCC	Ranks PC third of four	Degradation of environmental quality at PC (L_{11}).
CPUC+	For PC	Minimize supply interruption risk (C_{31}).
PC neighbors	Against PC	High seismic risk (L_{43}); degradation of environmental quality (L_{11}).
Sierra Club	Against PC	High seismic risk (L_{43}); degradation of environmental quality (L_{11}).

IV CONCLUSION

The FERC and the CPUC will consider more seismic risk data. ⑯

Key decisions

(4) *CCC* ranks PC third of four. ⑩
Procedure: non-judicial/open hearings; vote by 12 lay commissioners.
Key trade-off: environmental quality and seismic risk outweigh supply interruption risk.

(5) *CPUC* selects PC, conditional on more seismic data, etc. ⑪
Procedure: quasi-judicial hearing.
Key trade-off: supply interruption risk outweighs environmental quality and seismic risk.

(6) *ERA* approves import. ⑬
Procedure: open and closed hearing, deliberation.
Key trade-off: acceptable supply interruption risk due to earthquakes.

(7) *FERC* approves PC. ⑭
Procedure: quasi-judicial hearing.
Key trade-off: accepts loss in environmental quality (land use).

(8) *Washington DC Court of Appeals* remands case to FERC to consider seismic data. ⑮
Procedure: judicial hearing upon petition.
Key trade-offs: more seismic data over supply interruption risk due to need caused by delay.

Table 6.3. *(continued)* The MAMP framework applied to the decision process on LNG in the US.

(*d*) ROUND D: April 1981—end of 1982.

I PROBLEM FORMULATION

Assumptions:	(1) State and federal processes have selected Point Conception subject to acceptable seismic risk conditions. (2) FERC ruling on Point Conception will withstand NEPA best alternative challenge.
Questions:	Is Point Conception seismically safe? If so, should the site be given final approval in view of the decreased demand for natural gas?
Process constraints:	If Point Conception is not approved, the federal government can reopen the process only at appreciable costs.

II INITIATION

The CPUC and the federal court call for continuing deliberations considering new seismic risk information. (15)

III CONCLUSIONS

At the time of writing, the LNG Seismic Review Panel has judged the site to be seismically safe (16) and the CPUC and the FERC have conducted hearings concerning seismic risk at Point Conception. However, the utilities have announced their decision to defer construction of the terminal until an unspecified date since California no longer needs to import natural gas. (17)

Two key decisions were made during round A, and these are described in Table 6.3(*a*). First, the CCC implied that it could favor Point Conception over the nonremote sites due to concerns over population risk. This decision was made after long and difficult deliberations within the CCC, since its mandate was to protect the coastline and not to reduce the risks to residents of urban areas. This decision shows the difficulty of justifying trade-offs that are made, which explicitly expose people to safety risks. Against its own interests, the CCC thus advised Western to pursue at least one site in a remote area since it could deny approval to any nonremote site.

Secondly, the FERC indicated that it would not approve the Port of Los Angeles site because a recently discovered earthquake fault presented an unacceptable seismic risk, although no clear definition was given of what constituted an unacceptable risk. Contrary to many other siting issues, Los Angeles City Council, after considering this seismic risk, remained in favor of the terminal in the port. This turnabout, where the local interests favored a site and the national interests were opposed, shows the importance of political and economic factors. The Los Angeles city councilors found the potential economic benefits to the harbor, which was badly in need of new industrial development, more salient, especially in light of their short-term political perspective where they perceived their re-election to depend primarily on economic considerations. The federal authorities, on the other hand, would have found it difficult to justify siting the terminal in Los Angeles, since other, less risky alternatives existed.

A very different situation arose in Oxnard. The federal authorities could justify their approval of the Oxnard site based upon their internal risk assessment (FERC), which showed the probabilities of an accident to be justifiably low. But the economic situation in Oxnard contrasted sharply with that in Los Angeles in that the city was less in need of the project, and a middle-class housing development was located in the vicinity of the proposed site. The risk assessment commissioned by Oxnard City Council (SES) was, in this climate, decisive. The worst-case scenarios, showing that 70 000 Oxnard citizens could be victims of an LNG vapor cloud, provided the necessary arguments for those opposed to the terminal. However, the project never actually came to a vote in the council, since pressure emanated from other directions to change the siting procedures.

It was clear to all the parties involved that it was difficult, if not impossible, for the applicant to gain approval for a site under the existing siting procedure in California. In particular, there were possibilities of vetoing proposals at either the federal, state, or local levels, as evidenced by the respective reactions to the three proposed sites. In such a situation, as pointed out by Majone (1979), rather than trying to operate within existing constraints of the process, interested parties frequently try to change the rules of the game. Thus, the perceived stalemate that characterized the interactions in round A set the stage for what can be seen as a second round of interactions.

This behavior relates to the process described by Braybrooke (1978) who points out that issues are frequently transformed over time; round B is a good illustration of this process. The problem was redefined into two new questions: How should need for LNG be determined? If need can be established, what process would facilitate the identification of an appropriate site? Round B was thus initiated when pressure to change the existing siting procedures was brought to the California state legislature by the utility companies, the business community, and labor unions. Table 6.3(*b*) details the relevant alternatives that formed the basis for the debate on the elements of the proposed legislation, SB1081.

Round B contrasts with the others insofar as there was a single arena, the state legislature, compared with the array of hearings, statements, and so on, that characterized the interactions of the other three rounds. The debate can be pictured here as one between the pro-development interests personified by Senator Alquist, and the pro-environment and safety interests represented by Assemblyman Goggins, although, of course, the issue was not so clearly defined. A great deal of lobbying, where there were inevitably "hidden agendas", preceded the debate in the senate. A notable feature of this round was that the Sierra Club, which opposed land-based remote siting, was not part of the final compromise process.

The passage of the LNG Terminal Siting Act opened up a new procedure for finding an acceptable site, and thus set the constraints for yet another round of discussions (see Table 6.3*c*) with the following problem formulation: Which site is appropriate? Round C was initiated when Western filed for the Point Conception site.

Round C is a remarkable example of a democratic process for siting a technological facility. The CCC held numerous informal meetings with residents along the coastline and solicited suggestions from the citizens of California. Although the stipulation that the site submitted by the utility be included in this ranking, which was then passed on to the CPUC, can be questioned, this process stands as a good example of a participatory model for site selection.

As shown in Table 6.3(*d*), the initiating proposal in round D was determined by the activities in round C, which framed the alternatives as simply whether or not to declare the Point Conception site seismically safe. In addition, the reopening of the hearing procedures permitted the state and federal authorities to re-address the need question. Only two parties — FERC and CPUC — are currently active in the process. The LNG Seismic Review Panel has found that a terminal can be operated safely at the site, but the utilities have deferred pursing their application in the near future.

The MAMP framework illustrates the sequential aspect of the LNG siting process in California. The key decisions that were made, and the trade-offs that were explicitly or implicitly set by those decisions, are listed at the conclusion of each round in Table 6.3, showing how particular decisions constrained possible future options. During the course of the decision process, the need for imported natural gas in California diminished greatly. Instead of examining this need, the interested parties, "locked in" by previous decisions, examined the seismic data at a rather slow pace. This is an example of a process of non-decision making where the interested parties may be using existing political institutions and procedures to limit the scope of actions.

A second example of undesirable effects from sequential constraints concerns the planned and actual implications of the project on the supply of natural gas. Initially, the applications were for three separate sites, each with large storage capacities. During the course of the decision process, the number of sites was reduced from three to one, and the number of storage tanks at that site was reduced from four to two. Because of this concentration in one small area with less storage capacity than planned, the possibility of nondelivery resulting from bad weather or accidents would reduce the reliability of supply below the level Western intended in the original application. This transformation was not the result of a systematic analysis by Western, but was brought about by the sequential nature of the decision process.

THE USE OF RISK ANALYSES

One of the most striking features of the US decision process was the dominant role played by concern over public safety risk. As seen in Table 6.2, for all but two of the main parties population risk was a primary concern. In addition, this concern (including seismic risk) was used as a primary argument in 15 of the 22 times parties were listed in the interaction sections of Table 6.3 (rounds A—C). Finally, round D was focused almost entirely on seismic risk.

During the course of the LNG debate in California, six studies assessing the safety risks of the proposed terminals were conducted by the utility and local, state and federal government agencies. Several studies are of particular interest. Western, the applicant, commissioned a consulting firm, Science Applications, Inc. (SAI), to assess the risks of each site, and the FPC (FERC) produced its own risk analysis. These reports showed very low numbers on various probabilistic measures of risk (expected fatalities per year and individual probability of fatality per year), and these were interpreted to mean that the risk was acceptable. A risk analysis for the Oxnard site produced by the consulting firm Socio-Economic Systems, Inc. (SES) for the Oxnard municipal authorities suggested similarly low probabilistic measures of risk (though expected fatalities were 380 times higher than the SAI assessment). The SES study interpreted the uncertainty concerning safety to mean that it was high enough to preclude any confident statement that the facility posed a low probability of accident, and pointed out that the possible consequences of an accident could be catastrophic.

The formats for presenting the results differed markedly among the reports. The SAI study described maximum credible accidents in tables with accompanying probabilities. The SES study described worst-case scenarios in the form of graphic maps, with shaded areas showing where fatalities were likely to occur, but did not provide probabilities. Opposition groups interpreted the latter results as evidence that the terminal was *not* acceptably safe. The Oxnard City Council, originally in favor of the site, began to waver in its support, probably influenced by the apparent uncertainty of the risk and the strength of the opposition groups (Ahern 1980). In sum, the risk assessments did not provide a single, coherent assessment of acceptability of the risk of an LNG terminal; their results were subject to interpretation depending on party positions (Lathrop 1980). In fact, risk assessments were used both to promote and to oppose the terminal applications.

In reviewing the technical differences between the assessments, Lathrop and Linnerooth (1982) have shown that there are many degrees of freedom left to engineering and analytical judgment, including how to characterize risk, what formats to use for presentation, what gaps to fill with assumptions, which of several conflicting models to use, how to portray the degree of confidence in the results, and what contingencies simply to leave out of the analysis. This analytical freedom helps to explain the differences between the risk assessments of LNG discussed above, since it can push the risk measurement in any direction. For example, very conservative assumptions can drive it up; omissions of inconvenient aspects such as terrorism can drive it down. Clear presentations of expert disagreements can decrease the confidence in the results. The conclusions may therefore have as much to do with the predilections of the analyst as with the physical characteristics of the site or the technology.

This finding takes on special significance when viewed in the context of the policy process. Not unexpectedly, the risk assessments in this case study, though intended to advise clients on the safety of a proposed terminal, were, almost without exception, used at some time to support a party argument. For this reason, in many cases, clear incentives existed for the analysts to present their results as persuasively as possible. This explains the tendency on their part to omit discussions on the uncertainty of their results and to choose suitable presentation formats for making certain points regarding the LNG hazard.

In addition, the sequential nature of existing decision procedures limits the possibilities for comprehensive analyses, so that risk studies have no broader systematic evaluation into which to feed. They either address a narrow problem floating in an ill-defined broader problem, or they address a narrow problem after other variables have been constrained by previous decisions. In particular, the risk studies were carried out, not as an input to a broad energy siting analysis in California, but to support the problem of whether site *X* or site *Y* should be approved. Since round A in California was *not* formulated in these narrow terms (the question of whether the terminal was needed had yet to be resolved), the analyses were ill-suited to address fully the issues on the table. Analyses designed to address the question of safety were prematurely introduced into a process that had not resolved higher-order questions of energy policy. Although risk analyses could not offer (nor were they intended to offer) a panacea for the resolution of the siting question, they did serve to focus the debate on the safety question.[5] In contrast, the seismic risk studies of round D addressed the central question of that round, but only after most other variables had been constrained in previous rounds, again without a comprehensive evaluation.

CONCLUDING REMARKS

In the discussion of siting issues, it is important to distinguish between the questions *whether* to site the facility, and that of *where* to site it. Whether to have a facility at all ultimately depends upon national (regional) interests or objectives. In the energy debate, the lines are often drawn between two different objectives: one of large-scale technology, high economic growth rates, and a centralized level of decision making; or one of small-scale technology using, wherever possible, renewable resources or recycling, with a no-growth economy, and decentralized decision making. The resolution of the conflict between these objectives will depend on the political system, where national goals are arrived at through the interactions of various interested parties. In the case of the California LNG controversy, these parties included the industry or utilities, federal, state, and local governments, organized action groups, the unorganized consumers, those who benefit from an unspoiled coastline, and those who benefit, as well as those who do not, from *generalized* economic growth. The question of how parties view the political decision process is a basic issue that is germane to most of what follows.

An important consideration involves the *direction* of the decision process. In the US energy sector most projects are initiated by the industry, in contrast with the transportation sector, for example, where projects (roads, etc.) are planned by the government and are carried out by private industry after competitive bidding. A desirable mix of public and private enterprise involves trade-offs between the advantages of private initiative and those of national planning. In the case of energy supply, it is important to question whether broader national objectives can be achieved in the absence of a coherent national energy plan.

Where a project is conceived by industry before it has been considered by federal planners, there exists the danger of limiting further decisions to small variations of the proposed project. In the California case, one might ask whether this process, either before or after the 1977 siting legislation, encouraged an imaginative consideration of all the possible alternatives. For instance, was it necessary to have one large facility, or could there have been a series of smaller LNG storage facilities that would present correspondingly smaller risks on the same order as those of peakshaving plants located in industrial areas? Also, were the possibilities for offshore siting given sufficient consideration?

More specific to the issue of the direction of the precision procedures is one-stop licensing. Before 1977, it was necessary for the industry to obtain permits from many local authorities. However, this picture was changed by the 1977 LNG Terminal Siting Act, which gave one agency — the CPUC — the responsibility for granting a siting permit (it was also necessary to have federal approval, but the FERC appeared willing to accept the state's decision). An LNG facility brings prospects of regional economic development and may increase the tax revenues of the municipality, but it also imposes risks on local residents. For this reason, a procedure requiring local approval may prove difficult. Yet, if citizen participation in decision procedures is important, any process that is less sensitive to local preferences may be undesirable. Here one might investigate other mechanisms through which the preferences of the local population could be more fairly taken into account, such as "bidding schemes" that would allow for compensation to those who perceive themselves to be at risk, or that would enable municipalities to bargain over a siting decision. Such proposals are discussed in detail in Chapter 9.

In California the move to one-stop licensing represented a trade-off between public participation and licensing expediency. The purpose of the 1977 LNG Terminal Siting Act was to ensure a decision on an LNG terminal without extensive delay — at some sacrifice to local participation. This Act, in the interest of maximizing public safety and minimizing further delay, might have precluded finding an "efficient" solution to the siting decision by imposing the population density and the land-based siting constraints. By "efficient" we are asking whether, in the absence of these constraints, a site could have been identified that would have been viewed as more desirable by *all* the parties. One of the anomalies of the California siting process was that at one time nearly all the parties (including the utilities, the Sierra Club,[6] the CPUC, and the FERC) favored the Oxnard site, and yet it was ruled out by the state legislature primarily on the basis of a risk assessment showing that an accident at the site could result in catastrophic consequences.

The catastrophic dimension of the risks of LNG might therefore be viewed as having been the decisive element in the California LNG siting debate. The above hypothesis suggests that the facility would have been located at Oxnard, as recommended by the FPC (FERC), had it not been for the SES worst-case scenario showing that Oxnard could be engulfed by an LNG vapor cloud. After publication of this report almost all the relevant government agencies, as well as the public, became increasingly risk averse. A staff member of the California state legislature was quoted as saying that a terminal that could kill so many people *could not* be sited. The SES report also seems to have had a considerably greater effect on sensitizing the public to the risks of LNG than an earlier event — the explosion of an oil tanker in Los Angeles harbor. The latter showed that an accident could happen, whereas the former showed that a holocaust was possible!

Attributing the California LNG Terminal Siting Act, which required a remote site for the terminal, to the SES report showing worst-case scenarios should, however, be done with caution. This Act was the outcome of several complex factors, including industry's long-term aspirations to have a more expedient siting process that was not so heavily dependent on local politics, and the desires of at least one state legislator — Assemblyman Goggins — to build his reputation on opposing large-scale hazardous facilities. Both interests were satisfied by the Act with little or no cost to industry. The extra costs of a remote site are thus borne by the unorganized and, for the most part, unaware California consumers, and not by industry. Indeed, regulatory practice allows industry a profit based on a fixed percentage of its capital costs — and these costs are increased by the added expense of a remote site.

A comparison of the costs of the decision process for land-based sites versus the costs of more expensive, but in some ways, less risky, offshore alternatives was not considered during the California siting process. Since the costs of "delay" due to the time required to construct an offshore facility were considered important, it might have behooved the applicant and government decision makers alike to have asked: How would the final figures including the costs of delays due to procedural mandates and those due to design problems, compare for land-based sites and offshore sites?

In the course of the decision process, the need for natural gas in California diminished greatly, leading to the eventual withdrawal of the utilities from pursuing a terminal site at the present time. Yet, detailed and expensive investigations of the faults at Point Conception preceded re-examination of this need. The sequential nature of the decision process, in this case, did not adequately provide for the dynamic decision environment, where concerns such as the need for natural gas re-emerged and were not explicitly considered.

NOTES

1. An exception to this pattern is the California Coastal Commission, where the 12 lay commissioners preside directly at non-judicial hearings.
2. The circled numbers refer to the corresponding nodes in the PERT diagram in Figure 6.2.
3. The relationship between the federal and the state authorities for LNG facility approval and siting is deliberately vague. The federal government, by choosing not to clarify its mandate, has in effect chosen not to challenge state authority.
4. These sites, in order of their ranking, were: the US Marine Corps base at Camp Pendleton, Rattlesnake Canyon, Point Conception, and Deer Canyon.
5. It is therefore not surprising that round A ended in a stalemate.

Round B, where the state legislature took center stage, narrowed the problem (by resolving the question whether California needed a site) to one more receptive to technical risk studies.

6. The Sierra Club changed its stand in early 1977 to oppose the Oxnard site.

Chapter 7

LEG Risk Assessments: Experts Disagree*

One of the most challenging problems in decisions concerning the deployment of novel, large-scale technologies is the assessment of the risk to the surrounding populations. In particular cases, such as nuclear reactors or liquefied energy gas (LEG) facilities, the political process involved may tend to focus on one particular form of that risk: the risk to life from catastrophic accidents. This chapter examines several assessments of this type with two main goals in mind:

(1) to present and compare the various risk assessment procedures as they have been applied to LEG terminal siting, and in so doing to clarify the limits of knowledge and understanding of LEG risks;
(2) to quantify and compare the risks estimated in analyses prepared for four LEG sites, namely:
   - Wilhelmshaven (Brötz 1978; DGWE 1979; Krappinger 1978a,b,c; WSB 1978).
   - Eemshaven (TNO 1978).
   - Mossmorran–Braefoot Bay (Aberdour and Dalgety Bay Joint Action Group 1979, henceforth referred to as Aberdour; Cremer and Warner 1977; HSE 1978a).
   - Point Conception (ADL 1978; FERC 1978; SAI 1976).

The advantage of LEG technology, as explained in Chapter 1, is to reduce the temperature of a gas until it becomes a liquid, so that it can be transported and stored efficiently in tanks, with a high energy per unit volume. While LNG (mainly methane) is stored at $-161.5\,^{\circ}\mathrm{C}$ at very low pressure, LPG (mainly propane and butane) is stored at near atmospheric temperatures and pressures, leading to significantly different behaviors during spills. However, these three substances involve essentially the same

*This chapter was written by Christoph Mandl and John Lathrop.

accident scenarios, though with different parameters and probabilities of ignition.

Although there are many aspects involved in assessing the advantages and disadvantages of an LEG terminal at a specific site, the risk to the local population is one of the most crucial concerns. But because of a lack of historical data on accidents at LEG terminals, the frequency of such accidents as well as their consequences cannot be readily estimated. Over the past few years attempts have therefore been made to quantify the risk to local populations for different planned LEG terminals, using various techniques and models, with different results.

This chapter reviews the risk assessments undertaken for LEG terminals in four countries, discusses their plausibilities, explains their differences, compares their risk estimates, and draws conclusions as to their usefulness and limitations. Where necessary and appropriate, we also describe some of the reports in detail. While this is not the first comparison of LEG risk assessments (see, e.g., SES 1977) it is, to our knowledge, the first comparison of assessments from four countries.

Because LEG terminal risk assessment is a new technique, there is still disagreement between experts concerning how to quantify the risks, which models to use, what to include, and what to exclude. Clearly, no pretense is made that this chapter provides complete or final answers concerning comparative risks or risk assessments; rather, it describes some initial attempts to address important problems in the field of risk assessment.

RISK AND RISK ASSESSMENT

Before we can quantify risk, we must define it. It will become apparent in this section that different people mean different things when they talk about risk. Therefore our definition (actually a set of definitions) cannot be descriptive, but rather will be prescriptive.

Ideally, if one adopts the axioms of rational choice under uncertainty, the evaluation of any decision alternative should consider the probability distribution of the consequences of that alternative (see, e.g., Luce and Raiffa 1957). Yet the concept of risk singles out a subset of those consequences for special analysis. The term is typically applied to particular uncertain consequences, diverting attention from other costs and benefits that could be just as important in the evaluation. In the case of LEG, for example, several dimensions are of concern in site selection and facility design, such as cost, land use, environmental quality, air quality benefits of LEG, and dependence on foreign supplies. Some of these involve uncertain costs, such as financial losses to the developer if anything goes wrong (e.g., delay in application approval, loss of supply contract, vessel accident); environmental effects due to accidents or even routine disruption; fatalities or injuries due to supply interruption (e.g., unemployment, or health effects in a severe winter). These uncertain costs can be and are referred to as risks, and they can be analyzed by techniques of risk assessment, yet the term risk assessment in the context of LEG typically refers only to assessments of fatalities due to accidents. That is the scope for all the reports reviewed in this chapter.

Definitions of risk

Let us start by quoting some definitions of risk from the literature:

> "Risk is the expected number of fatalities per year resulting from the consequences of an accidental event." (SAI)

> "Risk is the probability of an injurious or destructive event generated by a hazard, over a specified period of time." (Cremer and Warner)

> "Group risk is defined as the frequency at which certain numbers of acute fatalities are expected from a single accident. The risk to society as a whole is defined as 'the expected total numbers of acute fatalities per year resulting from accidental events in the system'." (Battelle)

Surveying the set of risk assessments reviewed in this chapter, one can identify two polar definitions of risk. One extreme definition (Cremer and Warner) considers only the probabilities of destructive events and does not look at the consequences of these events. Such an approach only makes sense for comparison or evaluation in the very limited case when all destructive events have equally valued consequences, and risk is defined as the probability that any one of the events would occur in a given time interval. It would be meaningless to label two facilities equally risky if they had equal probabilities of an accident, but if an accident at one facility would have much more serious consequences than an accident at the other.

At the other extreme, risk can and sometimes is viewed as the event with the most serious consequences. Again we would argue that focusing on this kind of risk is not meaningful because it omits the probability of an event. We find the definitions of risk given by Keeney *et al.* (1979) the best because they address the following different aspects of risk:

(i) *Risk of multiple fatalities*: the probability of exceeding specific numbers of fatalities per year.[1]

(ii) *Societal risk*: total expected fatalities per year. This is appropriate for particular types of analysis, such as cost—benefit or risk—benefit analyses, where social preference is assumed to be linear in terms of number of lives lost.

(iii) *Group risk*: the probability of an individual in a specific exposed group becoming a fatality per year. This could be used to determine in some sense how much of the risk is being borne by neighbors, campers, boaters, etc. This definition also allows separate determinations of occupational and nonoccupational risks, which are often treated quite differently in political and social processes.

(iv) *Individual risk*: the probability of an exposed individual becoming a fatality per year. This is simply an average over the group risks measured by the third definition. This measure is somewhat troublesome because it depends on how an exposed population is defined. If "exposed" means an individual probability of fatality of greater than 10^{-12} per year, the individual risk will be averaged over a small region around the facility. If, on the other hand, "exposed" is defined with a cut-off probability of 10^{-30} per year, the individual risk will be averaged over a much larger region, and so will be much lower. In spite of this shortcoming, individual risk is a measure that allows a convenient comparison between the measured risk and more routine risks the individual may face; e.g., risk due to smoking, driving, etc. While such comparisons do not fit into a decision or choice framework, they do provide readily understandable benchmarks for scaling the risk of a facility.

Risk assessment as a decision aid

Given the orientation of this chapter, it is easy to forget that a risk assessment is not an end unto itself, but is in fact only one element in the complex process of LEG facility siting and design. More importantly, a risk assessment should be an aid for one or more of the decisions that must be made. A knowledge of where risk assessments fit within an LEG siting and design process is essential to the understanding of their adequacy and usefulness as decision aids.

Given the many decisions that must be made involving risk, there are several roles for risk assessment in LEG facility siting. Yet the processes studied in our research have narrowed that role down to a single application: on one dimension, risk to life and limb; and at one level, siting or design (depending on the country). There are several effects of this narrowing. First, it diverts analytical effort and political attention away from those questions not addressed by a risk assessment; for example, supply interruption risk could be a significant factor. On the other hand, it allows the analysis to be done. A fully comprehensive risk analysis might take decades, and would probably not be complete anyway.

A second effect of the narrow role given to risk assessment is that the level at which it is applied affects how it is conducted. When a risk assessment is part of the site selection process, a particular facility design is assumed, and analytical effort concentrates on such things as shipping traffic and local population density as site-specific inputs in a calculation of population risk. When a risk assessment is part of the facility design process the site is assumed fixed, and the analysis considers the sizes, arrangements, and specifications of components of the facility. In that case technical design variations are considered in terms of incremental risk reductions.

There is a third effect of the narrow role given to risk assessment. Once a site is selected, given the political realities of the situation, the question of the overall acceptability of the risk is more or less settled. If a risk assessment is applied at the design level, it may consider various modifications to reduce the risk in the most cost-effective way. However, given its scope and charter, the assessment is unlikely to find that the site cannot be made acceptably safe with current technology and to recommend that it should be abandoned. On the other hand, if a risk assessment is applied at the site selection level, it would at least be feasible to rule that none of the sites in the current choice set is acceptable.

Risk assessment does not exist in a vacuum. It is a decision aid within a much larger process. Any understanding of current methods of assessment, and any suggestions for improvement, also requires an understanding of that larger process. As this section has pointed out, that larger process controls the role and nature of risk assessments in very basic and important ways, even though they may be carried out as strictly independent studies.

REVIEW OF RISK ASSESSMENTS

In Table 7.1 we give a comprehensive overview of the most important risk assessment reports prepared for Wilhelmshaven, Eemshaven, Mossmorran–Braefoot Bay, and Point Conception. Some comments on the row headings of this table might be helpful.

(*a*) *Parts of the system considered.* Not all reports consider all the components of an LEG terminal system, namely: vessel, transfer, and storage tanks. In particular, for Wilhelmshaven there were two types of reports: one deals only with vessel operation and LEG transfer, the other deals only with the storage tanks.

(*b*) *Concept of risk.* As discussed above, there is no unique definition of risk. We have indicated the type of risk analyzed in each report.

(*c*) *Estimation of probabilities of events.* One crucial part of a risk assessment is the esitmate of probabilities, unless only the consequences are considered. It is therefore necessary to see how this problem has been solved in different reports. Two techniques can assist in performing this task for specific plants. The *event tree* is a technique for identifying a logical sequence of events (failures) that could result in unwanted consequences (accidents). Having identified the possible events, *fault tree analysis* attempts to determine the probability of a "top-level event" (typically a specific accident) that is the result of a sequence of events (failures) of the system. However, these techniques are not appropriate for estimating accident probabilities such as ship collisions. Two methods for estimating these probabilities are discussed later.

(*d*) *Estimation of consequences of events.* The consequences should be stated in terms a decision maker can understand. For this reason, and because of the definitions of risk typically adopted, most reports estimate the consequences in terms of the number of fatalities a certain event could cause.

(*e*) *Estimation of risk.* Different estimates are given depending on the definition of risk employed; in some cases no estimate is given at all.

(*f*) *Final findings.* As we see it, the ideal result of a risk assessment report is the quantification of the risk (in this case, from LEG) in comparison with risks from other sources. The ideal, and thus most useful, comparison is between risks from alternatives actually faced in the decision-making process; e.g., site A versus site B, site A versus no site, risk mitigation I versus risk mitigation II, etc. In any case, it should be kept in mind that decisions concerning the acceptability of the risk from LEG involve social value trade-offs and perhaps political considerations that go beyond the scope of the risk assessment and the legitimate authority of technical risk analysts. It follows that the final findings of a risk assessment should impart information to enable the decision maker to use them as a basis for a decision, without that decision actually being made.

(*g*) *Uncertainties in final findings.* Due to the limited data from LEG accidents there remains a substantial amount of uncertainty about the accuracy of the estimates of probabilities and consequences of such events. The reports handle this problem differently: some ignore uncertainties completely; some give conservative estimates; some perform sensitivity analysis; and some give error bounds on the quantified risk.

(*h*) *Single event with the highest risk.* If mitigating measures to reduce risk are to be undertaken it is helpful to know which event presents the highest risk. It is often the case that the highest-risk event offers the most cost-effective opportunities for mitigation.

When evaluating the risk assessment reports one should keep in mind that their differences, as shown in Table 7.1, can at least partially be explained by the fact that they were prepared and used for different decision processes. Each one was developed in a way suited to the particular process it was to serve.

ASSESSMENT AND COMPARISON OF LEG TERMINAL RISK

In this section we discuss the probabilities and consequences of different events (failures). First we consider the estimation of probabilities of failures, then the estimation of the size of the resulting vapor cloud and its ignition probability, and finally we consider the consequences to the local population. Our purpose is to compare the results of the risk assessment reports and to discuss important differences in estimates of probabilities

Table 7.1. Comparison of risk assessment reports prepared for the four sites.

Issues		TNO	Aberdour	Cremer and Warner	ADL
(*a*)	Parts of the system considered	Vessel, transfer, storage tank	Vessel	Vessel, transfer, storage tank	Vessel, transfer, storage tank
(*b*)	Concept of risk	Risk of multiple fatalities and and group risk	Group and individual risk	Probability of an injurious or destructive event	Multiple fatalities risk
(*c*)	Estimation of:				
	• probabilities of events	Yes, quantitative	Yes, quantitative	Only in terms of low, very low, etc.	Yes, quantitative
	• event tree analysis used	Yes	No	No	Yes
	• fault tree analysis used	No	No	No	Yes
(*d*)	Estimation of consequences of events	Yes, quantitative in terms of fatalities	Yes, quantitative in terms of fatalities	Yes, but only physical cons. (eg, spill size); no estimation of fatalities	Yes, quantitative in terms of fatalities
(*e*)	Estimation of risk	Societal and individual risk low cf. other man-made risks	Individual risk high cf. other man-made risks	No estimation of expressed fatalities; only of probabilities of events	Yes, quantiative
(*f*)	Final findings	Societal and individual risk low cf. other man-made risks	Individual risk high cf. other man-made risks	"No reason to doubt that installations cannot be built and operated in such a manner as to be acceptable in terms of community safety"	Point Conception suitable with respect to vessel traffic safety. Risk is very low.
(*g*)	Uncertainties in final findings	Not mentioned	Not mentioned	Not mentioned	Sensitivity analysis
(*h*)	Single event with highest risk	Grounding of LNG tankers	Not identified	Not identified	Not identified

Table 7.1. (*continued*)

FERC	SAI	Brötz	Krappinger	WSB
Vessel	Vessel, transfer, storage tank	Transfer, storage tank	Vessel	Vessel
Societal, group, and individual risk	Risk of multiple fatalities, group & individual risk	Not defined	Not defined	Not defined
Yes, quantitative	Yes, quantitative	Only in terms of very low	Yes,	Only in terms of very low
Yes	Yes	No	No	No
No	Yes	No	No	No
Yes, quantitative in terms of fatalities	Yes, quantitative in terms of fatalities	Yes, but only physical cons. (eg. spill size); no estimation of fatalities	No estimation given	Some quantitative statements in terms of few and many fatalities
Yes, quantitative	Yes quantitative	No estimation given	No estimation given	Yes, quantitative
Risk comparable to risks from natural events & thus on an acceptable level	"The risk is extremely low"	With regard to consequences & their probability there is no danger, cf. relevant laws	No final findings	Risk is not insignificant
Disagreement between experts is mentioned	Sensitivity analysis	Not mentioned	Not mentioned	Mentioned
Not identified	Not identified	Not identitied	Not identified	Not identified

and consequences in terms of the underlying assumptions of the models used and their plausibility. However, not all the reports are easily comparable. Some do not consider all the events we will be discussing, while others do not quantify either the probabilities or the consequences of events. Therefore, this section cannot be a complete comparison for all events.

In Table 7.2 we give brief descriptions of the planned terminals at Eemshaven, Mossmorran—Braefoot Bay, Point Conception, and Wilhelmshaven. Mossmorran is different from the others in that not only is this an export terminal, but the exported gases are LPG (mainly propane and butane), while LNG consists mostly (approximately 90%) of methane. As far as one can tell from the available risk studies, the layouts of the four terminals, the LEG tankers (except in size), storage tanks, and transfer systems are basically similar.

Probabilities of LEG spills

One of the most difficult questions in risk assessment is the identification of possible events or failures and the estimation of their frequencies or probabilities. By definition it is nearly impossible to get enough historical data to estimate the probability of a low-frequency event. Rather, one has to build models and rely on data from other, presumably similar, systems. Another important part of the risk assessment problem is the identification of events that have never occurred that could have serious consequences. This problem was acknowledged in the Lewis Report on nuclear safety (1978):

> It is conceptually impossible to be complete in a mathematical sense in the construction of event-trees and fault-trees; what matters is the approach to completeness and the ability to demonstrate with reasonable assurance that only small contributions are omitted. This inherent limitation means that any calculation using this methodology is always subject to revision and to doubt as to its completeness.

We therefore do not and cannot claim that the events considered here are a complete set of all possible events. However, the set includes all events that were accounted for in the risk assessment literature (e.g., TNO, SAI, ADL, Battelle).

The two major potential failures are vessel accidents and storage tank ruptures. Philipson (1978) describes two methods typically used to establish estimates of the probabilities of vessel accidents.

(1) *Statistical inference.* Estimates are computed using historical data, first for a larger class of ships, such as oil tankers, and then the estimates are modified to account for the anticipated differences in LEG ships and their operations at a specific harbor. This is done, for example, by employing judgment and by assessing the proportion of past accidents that would not have occurred if various mitigating measures had been taken.

Table 7.2. Descriptions of the four terminals and the sites.

	Eemshaven	Mossmorran–Braefoot Bay	Point Conception	Wilhelmshaven
	Import	Export	Import	Import
Type of transferred material	LNG	Propane/butane (liquefied) and gasoline	LNG	LNG consisting of 90% methane, 5% ethane, propane, and butane
Average transfer per day (in m^3 liquefied, or MW)	18 500 m^3 $\simeq$4 900 MW	13 400 m^3	Initial: 58 500 m^3 $\simeq$15 500 MW current plan: 41 000 m^3 $\simeq$10 900 MW	56 500 m^3 15 000 MW
Maximum capacity of ships	125 000 m^3	60 000 m^3 propane/butane 10 000 m^3 gasoline	130 000 m^3	125 000 m^3
Number of ships per year	54	80 for propane/butane 9 for gasoline	190	170 ships of 125 000 m^3 264 ships of 10 000 m^3
Number and capacity of storage tanks	2 × 120 000 m^3	4 × 60 000 m^3 propane/butane 2 × 31 000 m^3 gasoline	2 later 3 with 77 500 m^3 each	6 × 80 000 m^3
Number of people living within 2 km of terminal	60 (12/km^2)	approx. 350 (50/km^2)	projection for 1990: 14 (2.2/km^2)	0, but recreational area within distance
Number of people living within 5 km of terminal	858 (28.9/km^2)	approx. 8000 (200/km^2)	projection for 1990: 98 (2.5/km^2)	5900 (151/km^2)
Number of people living within 10 km of terminal	9800 (85/km^2)	approx. 100 000 (470/km^2)	data from 1977: 129 (0.9/km^2)	43 000 (275/km^2)

(2) *Kinematic modeling.* In SAI ship collisions are analyzed by assuming ship motions to be random in a zone of interest within the short time interval preceding an accident. A kinematic model provides the expected number of collisions per year for a harbor with specific configurations and traffic characteristics. A calibration to the actual conditions of a number of harbors is then made by scaling the model to fit actual past collision frequencies in these harbors.

Estimates of the probabilities of different types of events are given in Table 7.3. It should be mentioned that these estimates are not always taken directly from the reports; in some cases we have adjusted the estimates to take additional data into account. For example, SAI used more ships with larger tanks than those currently planned, so we reduced the probabilities and spill sizes accordingly. In FERC only spill sizes of 25 000 m^3 were considered, although data were also provided for smaller spill sizes, and these were considered in generating Table 7.3. The three Krappinger *Gutachten* (1978a, b, c) reported a variety of results using different accident reduction factors, ranging from 1.0 to 0.05. Because the latter factor was not based on any stated reasoning, we used the factor 1.0, which was used in Krappinger (1978a).

The most interesting findings from this comparison of assessments are:

(*a*) Compared with the probability of collision, grounding, and ramming, other events are considered to be rather unlikely (except for internal failure in Aberdour).

(*b*) The differences in spill probabilities between the three reports for Point Conception are substantial (between 10^{-3} and 10^{-6} for 10 000—25 000 m^3 spills).

(*c*) Although the traffic patterns at Eemshaven, Braefoot Bay, and Wilhelmshaven are quite different, they all come up with a total spill probability on the order of 10^{-3}. The spill sizes at Eemshaven and Braefoot Bay differ, and are not defined for Wilhelmshaven.

The event that could create the largest spill is the rupture of a storage tank, caused by severe winds, airplane or missile crash, meteorites, earthquakes, internal system failure, or accidents at chemical plants nearby. The estimate of TNO is taken from historical data from a peakshaving LNG plant. Cremer and Warner only qualify the probability as "remote", without reference as to how this qualification was produced. ADL and SAI derive their estimates from historical data on weather conditions, earthquake frequencies, and frequencies of airplane crashes. The probabilities for internal system (metallurgical) failure were derived from a technical analysis of the metal used and the temperature variations that could cause metal fatigue or stress. Using historical data from the FRG, Brötz estimates the probability of an airplane hitting one of the six tanks.

All LEG storage tanks are placed within containment basins capable of holding all the contents (in liquefied form) of the tanks. All credible failure scenarios assume that these basins will not break and therefore that all spills will be held. For common-cause failures (such as earthquakes, airplane crashes) this assumption is questionable. Only SAI considers the probability of more than one tank rupturing at a time due to a common cause. The maximum credible spill is then considered as a simultaneous rupture of all three storage tanks, each containing 77 500 m^3. SAI adjust their probabilities because the tanks are empty approximately 40% of the time.

Major findings on storage tank rupture probabilities are as follows:

(1) The probability of a storage tank rupture for all sites is estimated to be on the order of 10^{-5} per year.
(2) A conservative estimate of spill size is generally assumed to be at least the complete contents of one tank. However, Cremer and Warner assume that only 15% of the contents of one tank will be spilled.
(3) There are no major differences in the estimates, except between ADL and SAI. For example, the SAI probability of a spill due to objects crashing into the tank is 4×10^{-7}, while the ADL estimate is 10^{-5}. Elisabeth Drake (of ADL) has pointed out that this difference is due to changes in missile launch plans at the nearby Vandenburg Air Force base between the times the two reports were written (personal communication 1981).
(4) Common-cause failures that could cause more than one tank to rupture are only considered by SAI.

Consequences of LEG spills

We have discussed the probabilities of different spill sizes resulting from failures of parts of the system. Before we can quantify the number of fatalities that could result from these spills, we have to discuss what happens to the spilled LEG and how it can cause fatalities.

There seems to be agreement that only ignition and subsequent rapid burning or detonation of the spilled LEG can result in fatalities due to thermal radiation and blast effects. LEG will immediately start to vaporize after a spill, resulting in a vapor cloud, which, if not ignited, will then travel downwind and disperse. If there is no ignition, all parts of the cloud will eventually reach the lower flammability limit of concentration, below which ignition cannot occur. To estimate the effects it is therefore necessary to estimate the size of the vapor cloud, the downwind travel distance of the part of the cloud that retains a concentration above its lower flammability limit, and the probability of ignition.

Vaporization and dispersion. Among all topics of LEG risk assessment the question of how LEG behaves after a spill has attracted the most scientific interest. So far, empirical studies include only data for LNG

Table 7.3. Estimation of the probabilities of various events.

	TNO	Aberdour	ADL	FERC	SAI	Brötz	Krappinger
(1) Probability of collision that could lead to a spill per ship approaching the LEG terminal	2.8×10^{-5}			5×10^{-4}	1.3×10^{-8}	—	4×10^{-5}
(2) Probability of grounding that could lead to a spill per ship approaching the LEG terminal	2.5×10^{-4}	1.5×10^{-5} includes (2) and (3)		4×10^{-4}	0	—	7×10^{-5}
(3) Probability of ramming that could lead to a spill per ship approaching the LEG terminal	—			3×10^{-4}	0	—	3×10^{-7}
(4) Probability of missile or airplane crash causing one spill per year	—	—	See (14)	—	4×10^{-7}	8.3×10^{-5}	—
(5) Probability per year of a meteorite falling on a specific area of 1 m^2	—	—		—	3.3×10^{-13}	—	—
(6) Probability of internal system failure	—	3.2×10^{-3}		—	1.0×10^{-11}	—	—
(7) Number of ships per year	54	80	190	190	190	432	432
(8) Maximum size of ship (m^2)	12 000	6600	12 000	12 000	12 000	12 000	12 000
(9) Length of stay of loaded ship in the vicinity of the terminal (years)	—	—	2×10^{-3}	2×10^{-3}	2×10^{-3}	2×10^{-3}	2×10^{-3}
(10) Maximum size of one ship tank (m^3)	25 000	12 000	25 000	25 000	25 000	25 000	25 000

(11) Probability of different spill sizes given (1)							
$0 \leqslant 1\,000\ m^3$	0	0		0.02	0		
$1\,000 \leqslant 10\,000\ m^3$	0	0		0.026	0		0.05
$10\,000 \leqslant 25\,000\ m^3$	0.56	0.25	See	0.023	0.22	—	spill size
$25\,000 \leqslant 50\,000\ m^3$	0.44	0	(14)	0	0.025		not defined
$50\,000 \leqslant 75\,000\ m^3$	0	0		0	0		
(12) Probability of different spill sizes given (2)							
$0 \leqslant 1\,000\ m^3$	0	—		0.0024	—		
$1\,000 \leqslant 10\,000\ m^3$	0.33	—		0.0057	—		0.009
$10\,000 \leqslant 25\,000\ m^3$	0	—		0.0039	—	—	spill size
$25\,000 \leqslant 50\,000\ m^3$	0	—		0	—		not defined
$50\,000 \leqslant 75\,000\ m^3$	0	—		0	—		
(13) Probability of different spill sizes given (3)							
$0 \leqslant 1\,000\ m^3$	—	—		0.0034	—		
$1\,000 \leqslant 10\,000\ m^3$	—	—	See	0.0065	—		0.1
$10\,000 \leqslant 25\,000\ m^3$	—	—	(14)	0	—	—	spill size
$25\,000 \leqslant 50\,000\ m^3$	—	—		0	—		not defined
$50\,000 \leqslant 75\,000\ m^3$	—	—		0	—		
(14) Total probability of different spill sizes per year[a]							
$0 \leqslant 1\,000\ m^3$	0	0	0	2.3×10^{-3}	0		
$1\,000 \leqslant 10\,000\ m^3$	4.5×10^{-3}	1.1×10^{-3}	0	3.3×10^{-3}	0		3.8×10^{-3}
$10\,000 \leqslant 25\,000\ m^3$	8×10^{-4}	0	7.4×10^{-5}	2.5×10^{-3}	8.9×10^{-7}	—	spill size
$25\,000 \leqslant 50\,000\ m^3$	7×10^{-4}	0	3.2×10^{-6}	0	9.9×10^{-8}		not defined
$50\,000 \leqslant 75\,000\ m^3$	0	0	6.5×10^{-9}	0	0		

[a] = [(1) (11) + (2) (12) + (3) (13) + (5) (8) (9)] (7) + (4) + (6)

spills of up to 50 m^3 on land, and up to 200 m^3 on water. The prediction of the behavior of large spills has been based on theoretical models, which are not easy to validate. Predictions differ for large spills but have produced good estimates of observed spills.

The predicted maxmum downwind travel distances of flammable LNG vapor clouds after spills on water taken from the different reports are listed in Table 7.4. LEG vapor cloud dispersion on land is likely to be faster over rough terrain than over water, except in the case of LPG vapor clouds that could accumulate in low-lying areas due to their high density. The differences between the reports are substantial. While SAI and Brötz predict that such a cloud would travel relatively short distances, ADL and FERC predict large distances. It is also worth noting that the distance increases with decreasing wind speeds in FERC, while for SAI the distance decreases with lower wind speeds.

Although they are likely to be larger in size, spills on land are generally considered to be less dangerous than spills on water. Spills on land are usually confined because the storage tanks are surrounded by dikes, which are generally not expected to rupture. Also, the vaporization rate of LEG on land is slower than that on water.

Ignition of vapor clouds. A vapor cloud can be ignited by the event that caused the spill, or at a later time by some other means. As can be seen from Table 7.5, the probability of immediate ignition, depending on the event, is generally high since an event that can cause a tank to rupture can be assumed to create enough frictional heat to ignite the resulting vapor cloud. The probability that the vapor cloud will be ignited by some other means, given that it is not ignited immediately, depends on the presence of ignition sources within the flammable bounds of the cloud. Delayed ignition will in general have greater consequences because the cloud increases in size as it travels downwind. Therefore, for most spills immediate ignition will reduce the overall risk. In this respect, the estimates of TNO and Aberdour are more conservative than those of the other reports. Certainly, the immediate ignition probability can be site-dependent. For example, Keeney *et al.* (1979) point out that the immediate ignition probability at the particular site studied is high because collisions would involve large vessels carrying flammable cargoes. Because historical data on LNG spills are limited, estimated ignition probabilities cannot be validated.

The same model for delayed ignition probability is used by FERC, SAI, Battelle, and Keeney *et al.* (1979), who assume that each source of ignition has the same probability p of igniting a vapor cloud. Thus the probability P_n that the cloud will have been ignited by one of n sources becomes $P_n = 1 - (1 - p)^n$. Additionally, all the reports using this model assume that each person is a source of ignition, because he or she is likely to be using facilities (e.g., car, oven, light) that could ignite the vapor. The main

Table 7.4. Maximum downwind travel distance of LNG flammable vapor clouds following instantaneous spills on water.

Report	LEG spill size (m^3)	Atmospheric stability[a]	Wind speed (km/h)	Downwind distance (km)
Brötz	20 000	A–F	All wind speeds	2.3
		During night only	All wind speeds	3.5
TNO	25 000	D	—	3.3
		E, F	—	10.0
ADL	25 000	A	25.0	1.0
		D	21.0	7.0
		E	19.8	10.0
		F	10.8	20.0
FERC	30 000	A	25.0	0.5
			16.0	0.5
			9.0	0.6
		D	25.0	4.2
			16.0	4.9
			9.0	5.9
		E	25.0	7.8
			16.0	9.2
			9.0	11.3
		F	25.0	18.1
			16.0	21.6
			9.0	27.1
SAI	37 500	A, D, F	54.0	6.0
		A, D, F	25.0	3.5
		A, D, F	11.0	2.0
		A, D, F	0.0	1.0
ADL	50 000	A	25.0	1.0
		D	21.0	9.0
		E	19.8	15.0
		F	10.8	25.0
SAI	88 000	A, D, F	11.0	2.5

[a] Atmospheric stability: ranging from A, very unstable (rough) to F, very stable (calm).

differences between the reports lie in the judgmental estimates of the probability p, as shown in Table 7.6.

The assumed values of p can be viewed as either conservative or non-conservative depending on the number of people (and thus ignition sources) within the bounds of the vapor cloud. The estimate of FERC, for example, is less conservative for Point Conception than that of SAI because there would be less than 130 people living within a 10 km radius of the LNG facility. Thus the FERC estimate implies that there is a substantial probability that the vapor cloud will not be ignited at all, while that of SAI implies that the cloud will be ignited with very high probability.

Table 7.5. Probabilities of immediate ignition following vessel tank rupture caused by different events.

Event causing the ignition	TNO	Aberdour	FERC	SAI	Battelle	Keeney *et al.* (1979)
Collision	0.65	0.66	0.9	0.9	0.8	0.9–0.99
Grounding	0.1	—	0.0	—	0.3	—
Ramming	—	—	0.9	—	—	—
Missile/airplane	—	—	—	0.9	0.9	—
Meteorite	—	—	—	0.0	—	—
Internal failure	—	0.9	—	0.0	—	—

Table 7.6. Ignition probabilities per person in case of delayed ignition.

	FERC	SAI	Battelle	Keeney *et al.* (1979)
Probability p that each person within the vapor cloud ignites the cloud	0.01	0.1	0.01	0.01–0.1

On the other hand, using the model for Wilhelmshaven, with 43 000 people living with 10 km of the LNG site, the FERC estimate implies that the vapor cloud will be ignited, but only after covering a more densely populated area than that predicted using the SAI estimate.

Fatalities caused by ignited vapor clouds. Effects of ignited vapor clouds include thermal effects (e.g., fire, burning) and blast effects (e.g., mechanical destruction, shock waves). There is no doubt that thermal effects exist, but it is an open question whether blast effects due to a deflagration or detonation can occur at all with methane and, if so, whether the peak overpressure thus created is sufficient to cause damage. TNO considers that blast effects are the only serious danger and that thermal effects are comparatively small. Cremer and Warner consider both thermal and blast effects, since the Mossmorran terminal would handle butane, propane, and ethylene, which are known to explode in certain mixtures with air. ADL only considers thermal effects, because methane explosions (either deflagration or detonation) are believed to be very unlikely. FERC and SAI again only consider thermal effects, while Brötz considers both thermal and blast effects. In NMAB (1980) it is concluded that LNG vapor cloud explosions cannot be ruled out completely, even though there is no empirical evidence for such a possibility.

A first step in estimating the fatalities within certain distances from a vapor cloud is to determine the levels of thermal radiation and peak overpressure above which fatalities can be expected. Here one has to distinguish between primary and secondary effects. Primary effects are fatalities directly caused by thermal radiation and peak overpressure; while secondary effects are fatalities from fires caused by thermal radiation, and from collapsing buildings resulting from peak overpressure.

All reports consider only primary thermal effects and secondary blast effects. Brötz maintains that primary blast effects can be ruled out because the required peak overpressure has never been observed. Secondary thermal effects, however, may affect people sheltered from direct radiation, but are very difficult to estimate. One way to include them is to assume a low radiation threshold level for fatalities. Blast effects are not significant in the risk calculations of most of the reports studied. The only report dealing with these at some depth is TNO; Brötz does not consider them at all. The treatment of thermal effects also varies markedly. The distance from the center of a fire to the lower fatality level is about twice as large in ADL than in FERC and SAI. Neither Cremer and Warner nor Brötz give a lower fatality level.

Summarizing the major findings of our comparison of fatality calculations we find:

(*a*) The reports disagree on the major causes of fatalities. While TNO assumes that all fatalities would be caused by secondary blast effects, ADL, FERC, and SAI assume that they would all be caused by thermal radiation. Neither Cremer and Warner nor Brötz consider fatalities from vapor cloud explosions or fires.

(*b*) There are disagreements as to the radiation level above which there will be fatalities. ADL adopts the most conservative estimate on this topic.

(*c*) The effects of LNG and LPG spills can be quite different. For example, it is known that LPG vapor can explode, while the possibility of an unconfined LNG vapor cloud explosion has not yet been determined.

(*d*) The ignition of an LEG vapor cloud is likely to affect nearby chemical plants, as well as the people living or working in the vicinity. With the exception of Point Conception, there are chemical plants near all the planned terminals. Both Cremer and Warner and Brötz consider this, but conclude that these effects would not increase the overall risk significantly. TNO points out that in the case of detonation a nearby ammonia storage tank in Rotterdam harbor could collapse with disastrous consequences (a lethal cloud of ammonia would extend tens of kilometers).

Assessment of population risk

The estimates of the societal risk, the individual risk, and the risk of multiple fatalities are given in Table 7.7; no estimates were given by Cremer and Warner or by Brötz. It should be noted that the estimates of SAI were for an LNG terminal with more storage tanks and larger ships than the one currently proposed. Not surprisingly, Point Conception has

Table 7.7. Estimates of risks for the different sites.[a]

	TNO	Aberdour	ADL	FERC	SAI
Societal risk (fatalities per year)	4×10^{-2}	Not estimated	7×10^{-6}	10^{-5}	10^{-6}
Individual risk (probability of fatality per year)	$\leqslant 7 \times 10^{-6}$	7×10^{-4}	$\leqslant 9 \times 10^{-8}$	8×10^{-7}	10^{-8}
Number of people at risk	$\geqslant 5000$	Not defined	$\geqslant 80$	15	90
Risk of multiple fatalities: probability that number of fatalities per year is equal to or greater than:					
1	3×10^{-3}	Not estimated	10^{-6}	Not estimated	6×10^{-7}
10	10^{-3}		10^{-8}—6×10^{-7}		3×10^{-11}
100	5×10^{-6}		∅		∅
1000	5×10^{-6}		∅		∅
5000	3×10^{-7}		∅		∅

[a] Note that the SAI estimates have not been adjusted to make them compatible with ADL and FERC, as has been done elsewhere in the text. Therefore, the risk of the smaller LNG terminal currently planned, as estimated by SAI, would be lower than those presented here, so that the differences between risk assessments here and in the text are in some sense understated.

the lowest population risk of the three sites. However, differences in the estimates should be interpreted with caution, since they consider quite different events. The probabilities for the same event, even for the same site, also vary between different reports. A more subtle difference between the reports is that the total number of people exposed N is defined in different ways, as can be seen by comparing the three Point Conception reports in Table 7.7. This difference can be important because N determines, and in some sense even defines, the individual risk measure in a way that may not be clear to the reader of the report.

While the relative ranking of Point Conception is not particularly surprising, what is striking about the estimates in Table 7.7 is the magnitude of the differences. Societal risk, individual risk, and the risk of one or more fatalities vary over four orders of magnitude across sites, and the risk of ten or more fatalities varies over eight orders of magnitude across sites. It is hard to imagine another area of political concern where performance measures receiving as much attention as these did could vary over such a wide range. Yet even more striking are the differences between the three reports prepared for Point Conception. There is about a factor of ten difference in both societal and individual risk (but note that the differences in individual risk can be partially explained by the different numbers of people exposed). There is a difference of four orders of magnitude in the risk of ten or more fatalities. A policy maker faced with such variations could conclude that all three reports are based on very limited knowledge of the risks of LEG.

EVALUATION OF LEG RISK ASSESSMENTS

It has become clear that there are substantial disagreements on both the probabilities and the consequences of events, even for the same site. There also exist discrepancies in reports for different sites that cannot be explained solely by the LEG terminal and the respective locations. However, because risk assessments are produced for specific decision-oriented purposes we also have to ask whether or not they serve those purposes well. Thus we now evaluate the reports from two aspects: their scientific quality, and their usefulness in the decision-making process. Although these are certainly interrelated there are specific questions related to each one.

(1) Given the reports discussed above, can the scientific quality of risk assessments be improved, and in what way?
(2) Given a risk assessment, even a scientifically perfect one, can it be extended or improved to become a more effective aid in a decision-making process?

Scientific quality

The one clear objective of a risk assessment is to estimate the level of risk from a planned LEG terminal facility. There are three criteria related to the scientific quality of a risk assessment: the terms of the risk definition used; the accuracy of the estimate in some probabilistic sense; and the verification of its accuracy. The question of accuracy is crucial to the entire area of risk analysis. It was broadly discussed when the Rasmussen Report was first published (NRC 1975), and has remained an important issue ever since. As we have shown, there are substantial differences between the risk assessments examined here on three dimensions: the events considered, their probabilities, and their consequences. All reports claim that their estimates are conservatively high, and only a few mention uncertainties. We now discuss the different problems of accuracy in more detail.

(i) *Events considered.* The reports disagree substantially on the events considered, even for the same LEG facility. One possible event, sabotage, is not considered in any of the reports. It would be scientifically more accurate to say that the results of the risk assessments are conditional estimates of the risk, assuming that particular events, such as sabotage, do not happen. It is important to realize that this is a simplification of reality that any scientific model builder has to cope with. However, it is important to mention in a risk assessment the assumptions under which the estimate is valid. Clearly, certain events are not considered for good reasons, in particular because specific events are *a priori* felt to be unimportant or because they are nearly impossible to quantify, like sabotage. This results, however, in the notion of risk assessment as being an estimate of risk *conditional on certain assumptions.*

(ii) *Estimation of probabilities.* Probabilities are derived from frequency distributions, personal (expert) judgments, and combinations of the two. The use of historical data to estimate probabilities of future events is very appropriate, unless major changes in the likelihood of events occur. If future events are expected to differ from those in the past, or if historical data do not exist, one has to rely on the judgment of experts to estimate probabilities. This method — the Bayesian approach to probability — is conceptually well founded in the axioms of rational behavior (see Lindley 1973, Luce and Raiffa 1957) and is extensively used in decision analysis under a single important assumption: that a single decision maker is involved. In that case it is conceptually feasible to use the decision maker's personal judgment, or a single source of expertise commissioned to generate the probabilities. In our case, however, this assumption is not appropriate. If the estimation of risk depends heavily on expert judgment of probabilities, the results can be biased by the specific experts used. One set of results cannot be readily held to be more reliable than another set generated by an expert who believes in different probabilities. Yet in analyses of processes as poorly understood as LEG accident scenarios, expert judgment of probabilities cannot be avoided. While it is clearly desirable to minimize the role of such estimates (in favor of more objective probabilities), this requires data collection that is expensive in terms of both time and money and may be sometimes even impossible. In any reasonable LEG siting process, trade-offs will be made between cost and decision quality in such a way that expert judgement probabilities play an important role. It follows that in any LEG risk assessment, the sensitivity of the risk estimates to a range of different expert judgments should be clearly specified.

(iii) *Estimation of consequences.* Before the consequences of an event can be analyzed it must be clear what aspects of risk are to be considered — e.g., fatalities, injuries, or financial losses. To estimate one of these risks, the consequences must be stated in the same terms, i.e., the number of fatalities, the number of injuries, or the amount of financial losses. Reports that consider consequences only in terms of the amount of LEG spilled, density of thermal radiation, or the like, cannot estimate risk in terms the decision maker or the public can easily understand. Many, but not all, reports consider consequences in terms of the number of fatalities; none considers injuries or financial losses explicitly, and some consider the consequences only in terms of spill size or thermal radiation. Even those reports dealing with the number of fatalities differ to some extent with respect to consequences, mainly due to the lack of experience with large LEG spills and doubts as to how to model LEG vapor cloud behavior. Thus it is not known with certainty how many people are potentially at risk, although it is possible that these uncertainties could soon be reduced in the light of current experiments on large LEG spills.

Having presented the crucial issues in LEG risk assessment we turn now to the problem of evaluating the accuracy of such assessments. Although we have identified some important factors, it is nonetheless very difficult to identify the most accurate risk assessment because we are dealing with extremely rare (or low-probability) events. This calls for some explanation.

Crucial to the growth of scientific knowledge is the ability to marshall evidence that a certain scientific statement or theory is false. In fact, as some philosophers (such as Karl Popper) argue, it is not possible to prove the correctness of a scientific statement, but only to prove its falsity. An example from the history of physics supports this view: a very strongly held theory, Newton's laws of motion, was only shown to be false, or, to be more precise, was shown by Einstein to be a special case of relativity theory after more than 200 years. Not all scientific statements can be shown to be false, for two reasons. One is that a scientific statement may be so accurate that any attempt to prove its falsity fails; the other reason is that a scientific statement may be so imprecise, or concerns such inaccessible events (e.g., in the distant future), that it is not currently possible to prove its falsity. But the latter type of scientific statement is much weaker than the first. While the laws of physics are examples of the first type, Marx's forecast of the economic development of industrialized countries is a scientific theory of the second type.

Let us now look at LEG risk assessment. It is not difficult to design experiments that could prove that certain predicted consequences are false. This could be achieved, for example, by experimenting with large LEG spills. But it is difficult to prove that the risk estimate is incorrect because important events are not considered or because the probabilities are wrong. The reason for this difficulty is that, by definition, low-probability events occur very rarely. Even the data on LEG terminal system behavior for 20 or 30 years would provide insufficient information, because very few accidents, if any, would have occurred. Such experience can only put an upper bound on a probability of occurrence at a particular confidence level, a bound that may be far greater than the very low probabilities involved in LEG risk assessment.

There is no easy way to resolve the problem of validating estimates of low-probability events. For this reason, Weinberg (1982) considers risk assessment to be an "art" because "there are, and will always be, strong trans-scientific elements in risk assessment." However, the fact remains that risk assessment is an "art" that is a more desirable means of estimating risk than any other alternative.

Limitations of risk assessments

We will now discuss the shortcomings of LEG risk assessments as decision aids, and how they can be improved. It is first necessary to identify the types of decisions for which risk assessments are used. Basically two distinct decision problems are involved: the choice of a specific site for an LEG facility, and whether or not certain risk-mitigating measures should be introduced. Although it is not clearly mentioned in the reports, it seems that those prepared for Point Conception and Eemshaven were used as inputs to both types of decisions, while the reports prepared for Mossmorran--Braefoot Bay and Wilhelmshaven were primarily used for decisions on mitigating measures.

Of course, decision problems are not only concerned with risk; consequences such as costs, benefits, environmental impact, supply interruption, and risk to life are also important dimensions. From a decision analysis point of view one should first clarify the alternatives and then quantify the consequences for each alternative, as shown in the decision trees of Figure 7.1. The siting decision is of course not independent of the choice of which technical standards to employ for the terminal, a fact that further complicates the process. However, at some point a trade-off has to be made between the various consequences of the decision, e.g., costs, benefits, risk, etc. This is exactly the point where questions such as "How safe is safe enough?" or "What level of risk is acceptable?" are raised. In a decision analysis framework one should instead ask questions like: "Is it preferable to reduce the expected number of fatalities and improve the layout of the plant (with additional costs), or should the expected number of fatalities remain at 10^{-6} per year and the technical standard of the LEG facility also remain the same as planned?" There is no scientific approach to answering such questions. In fact, there is not even a unique way of answering them for society as a whole, because we do not know of a unique way to aggregate preferences of individuals into a societal preference list. No matter what the results of the risk assessment may be, different people may prefer different alternatives for the problems posed in Figure 7.1 because individual trade-offs between risk and other consequences are likely to differ.

In the past, decisions concerning technological risk have largely been made by the engineers and companies who plan the facilities: civil engineers decide on the safety level of dams, and airplane industries and

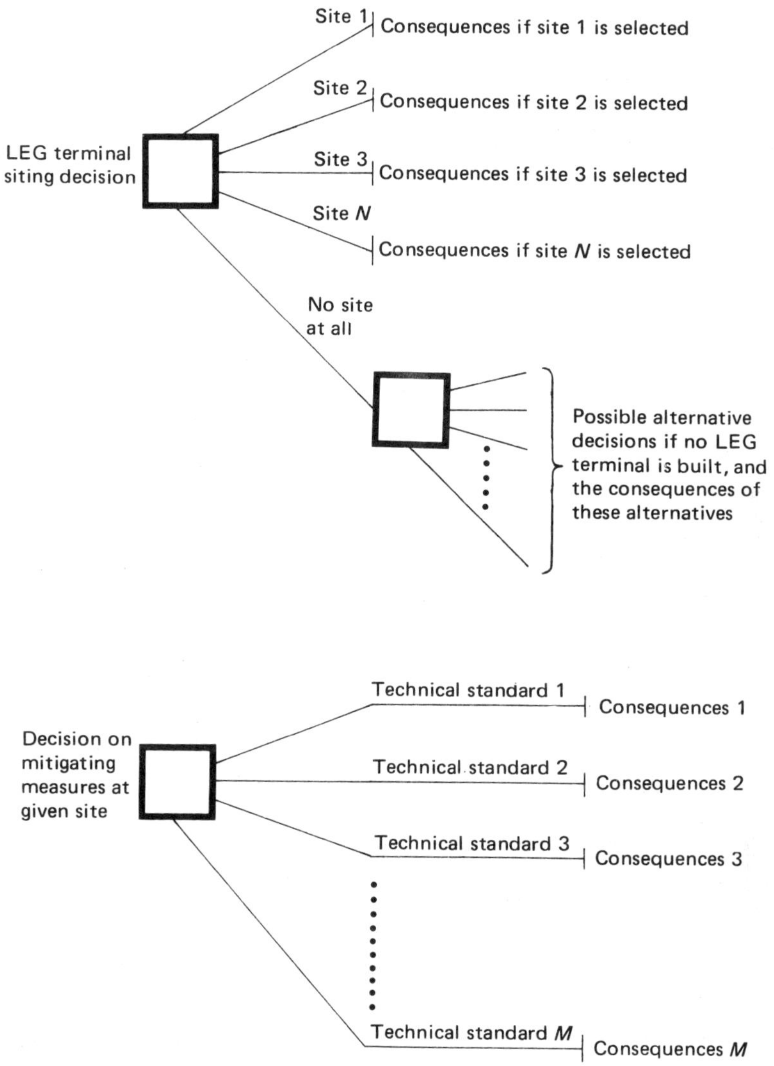

Figure 7.1. Decision trees for the main decisions on LEG terminals.

regulatory commissions decide on the safety level of airplanes. Although this is a generally accepted practice one must be aware that underlying these decisions are subjective trade-offs between safety (or risk) levels and other factors. Even if people or interest groups do not accept the subjective trade-offs of experts or engineers, there is no scientific way to prove that these experts are correct. As shown in Table 7.1, the Cremer and Warner and Brötz reports state that the risk levels were acceptable, a statement that cannot be defended on scientific grounds, although this does not mean, of course, that the risk level was unacceptable to all the parties involved.

It is also important to realize that the consequences of decisions will affect groups or individuals in different ways. Not many people in a society are exposed to the risk of LEG terminals, while many receive the benefits. On the other hand, if no individual is exposed to the risk because an LEG terminal is not built, the consumers might have to pay a higher price for natural gas. Thus there is a conflict of interests between different groups that cannot be solved by scientific means. Although our reasoning here is not new, there still seems to be a tendency within the political process to believe that risk assessments can make decisions more objective or that the people who do the risk assessments can also answer the question of whether or not the risk resulting from a specific decision is acceptable. Although risk assessments cannot solve the problem of people having different trade-offs with regard to the consequences of decisions related to LEG terminal facilities, risk assessment may still be useful. Its uses are discussed further in Chapter 8.

Differences between risk assessments

As discussed above there are important differences in risk assessment reports, especially between those for Wilhelmshaven and Mossmorran–Braefoot Bay on the one hand, and Point Conception and Eemshaven on the other. These discrepancies could be attributed to a number of factors, including differences in cultural styles, in the risk analysts responsible for the reports, in the purposes of the reports, in analytical and financial resources, or some other unknown factors. Although it cannot be substantiated, the expectations of the clients, as well as the scientific background of the analysts, are also likely to be significant reasons for the differences.

While the reasons for the differences cannot be ascertained, their extent indicates the great amount of leeway left to analytical judgment. As this chapter has made clear, several decisions must be made in the course of performing a risk assessment, such as how to characterize risk, what presentation formats to use, what gaps to fill with assumptions, what assumptions to adopt, which of several conflicting models to use, how to indicate the degree of confidence of the results, and which events simply to omit from the analysis. These decisions can push the results in any direction. Very conservative assumptions increase risk estimates; clear presentations of expert disagreements can reduce confidence in the results; and particular formats highlight particular aspects of the risk. Such factors can push the results of an analysis over such a large range that the final result may be affected more by the predilections of the analyst than the physical features of the site or technology. This same result was found in the comparison of three risk assessments performed for the proposed terminal at Oxnard (FPC, SAI, and SES; see Lathrop and Linnerooth 1982).

Table 7.8. Ranking of reports for Point Conception in terms of conservatism of estimates on specific issues.

	Conservative	Less conservative	
	1	2	3
Completeness of events considered	SAI	ADL	FERC
Spill probability due to ship collision	FERC	ADL	SAI
Spill sizes considered	ADL	SAI	FERC
Spill probability due to storage tank failure	SAI	ADL	–
Spill sizes considered	SAI	ADL	–
Maximum travel distance of flammable vapor cloud after spill on water	ADL	FERC	SAI
Maximum travel distance of flammable vapor cloud after spill on land	ADL	SAI	–
Lower fatality level of thermal radiation	ADL	FERC	SAI
Delayed ignition probability	SAI	FERC	–
Overall risk	FERC	ADL	SAI

When looking at the three reports for Point Conception, however, there is no indication that one report is more conservative (or less conservative) in its overall risk estimate than the others, as shown in Table 7.8.

Dealing with uncertainties

There are difficulties in determining and protraying the accuracy of risk estimates. Some reports deal with only a few of the uncertainties, as shown in Table 7.9, but the greatest uncertainties are acknowledged in the Battelle report:

> For both the number of expected fatalities per event and the corresponding annual frequencies, the lower limit of the confidence intervals should be considered to be significantly less than 0.1 times the given values, the upper limit over 10 times the given values. In some cases the upper limit is given by the total number of persons at risk.

The reports not listed in Table 7.9 do not explicitly consider uncertainties in the risk estimates at all; even those that do differ as to what these are, and the relevant range that these values can take. We consider as particularly uncertain the estimates of probabilities of events stemming from expert judgment. Therefore sensitivity analyses should involve at least the following parameters:

- probability of an LEG vessel accident resulting in a spill;
- probabilities of immediate and delayed ignition of an LEG vapor cloud;

Table 7.9. Uncertainties in risk estimates.

Report	Range of expected number of fatalities per year	Reasons for range
ADL	4×10^{-6} to 7×10^{-6}	Population density — present or future; active time of ignition source
SAI	1.2×10^{-6} to 1.201×10^{-6}	Percentage of fatalities among people enveloped by a burning plume, and other conservative assumptions
Battelle	At least 10^{-6} to 10^{-4}	Probability of vessel accident, probability of immediate and delayed ignition of LEG vapor cloud
Keeney *et al.* (1979)	1.7×10^{-5} to 2×10^{-5}	Probability of ignition per source, maximum travel distance of flammable vapor cloud
SES (1977)	1.5×10^{-2} to 5.7×10^{0}	Maximum travel distance of flammable vapor cloud, probability of LEG vessel accident

- probability of transfer system and storage tank failure resulting in a spill.

Another topic, the maximum downwind travel distance of an LEG vapor cloud, seems to us to be less critical because this will be primarily determined by the delayed ignition probability. This view is also supported by Battelle:

> The uncertainties resulting from the application of simple, experimentally unverified models describing the dynamics of the LEG vapor are not considered to be critical for the assessment of the risk.

To show how large the differences between accident probabilities can be, it is useful to examine the probability of a spill per year at Point Conception (Table 7.3). This probability was estimated at 9.9×10^{-7} by SAI, as 7.7×10^{-5} by ADL and as 8.1×10^{-3} by FERC. If the range on the order of 10^4 is taken as the possible probability, the range of the societal risk given in Table 7.7 should also be of the order of 10^4, because the probability of a spill on water is essentially related to the expected number of fatalities per year. The FERC and SAI risk estimates in Table 7.7 do not differ a great deal because the former makes less conservative assumptions on other issues, as shown in Table 7.8.

From this small example of sensitivity analyses we would argue that the range of uncertainty stated in Battelle and SES of the order of at least 10^2 is defendable as a minimum range for most risk estimates. Of course, this is a very rough estimate. It is certainly necessary to perform thorough sensitivity analyses to make more precise statements on the ranges of uncertainties.

But a more general point should be made here. Each report poses as a representation of the current state of knowledge regarding LEG risks, but because that knowledge is incomplete, some of the reports represent it using probabilistic terms or error bounds. Yet each report is based on a different state of knowledge: different assumptions are made, models used, probabilities estimated, etc. No one report in fact represents a comprehensive representation of the current state of knowledge. When SAI gives a probability of 9.9×10^{-7}, and FERC gives a probability of 8.1×10^{-3}, for the same event, the policy maker is likely to be somewhat at a loss as to the appropriate figure upon which to base his or her decisions. There is some "societal subjective probability" that most likely lies between those two probabilities, since each represents only a subset of the total state of knowledge. Yet neither report acknowledges that the other estimate exists! What is needed is a "meta-analysis" combining the different estimates and models, so that it represents a larger fraction of the available knowlege than any one of the existing risk assessments.

A meta-analysis would be more objective than any of the existing reports because if two consulting firms were each asked to do such a meta-analysis, the two reports would probably be more in agreement than, say, SAI and FERC. Such an analysis would also have much larger error bounds, or broader probability distributions than the existing analyses. While policy makers would prefer more precise estimates of risk, our present state of knowedge simply does not warrant such statements. The imprecision in our knowledge about LEG risks should be clearly communicated to readers of risk assessment reports.

GUIDELINES FOR STANDARDIZED RISK ASSESSMENTS

In this section we organize many of our findings in the form of guidelines for a standardized LEG terminal risk assessment. These are meant to suggest ways of improving assessments, and also to enable someone not familiar with the field to evaluate such reports. The guidelines are intended to improve both scientific and decision-aid aspects of LEG risk assessment. Much of what we present here has already been stated at some point in this chapter, yet we feel it is useful to present these issues in a summary form.

(1) *Definition of risk.* Because there are several different definitions and concepts of risk, the particular one used in an assessment should be made clear. In addition, the reason for the choice of that particular risk definition should be explained.

(2) *Completeness of considered events.* It is conceptually impossible to be sure that all possible hazardous events have been included in the assessment. However, the events listed in Table 7.3 can be considered

a minimal set to be included. Other events that could add substantially to the risk (such as sabotage), but were not considered for some reason, should also be mentioned. Thus the reader of a report should realize that the validity of the risk estimate is conditional on the consideration of only certain events.

(3) *Estimation of probabilities.* Whenever possible, probabilities should be estimated using data rather than judgment. Whenever judgmental probabilities are used they should be identified. Furthermore, a number of experts should provide estimates so that a range of possible judgmental probabilities is generated.

(4) *Estimation of consequences.* The consequences should be expressed in terms that concern the decision makers (e.g., fatalities, injuries, financial losses) rather than in physical terms (e.g., spill size, thermal radiation). The possible consequences of domino effects (between an LEG terminal and nearby chemical plants, for example) should also be considered. Whenever possible, consequences should be estimated using data from experiments rather than theoretical, unverified physical models.

(5) *Identification of system parts that present the maximum risk.* For considering mitigating measures and engineering design features, it is helpful to identify the parts of the system that present the greatest risk.

(6) *Sensitivity analysis.* Any risk assessment report should perform sensitivity analysis, particularly on the judgmental probabilities, to show the possible range of uncertainty of the risk estimate.

(7) *Assumptions.* The assumptions on which the analysis is based should be clearly stated. In addition, wherever possible the implications of each assumption should be presented to aid comparison with other assessments.

(8) *Risk—benefit analysis.* Although the estimation of the risk itself increases the understanding of the implications of certain decisions, we feel that the estimation of the risk and the benefits of alternatives in the context of LEG terminal siting and design problems would be more appropriate and useful to the decision makers.

(9) *Acceptable risk level.* There is no scientific way to decide whether a certain risk level is acceptable to society or not. Therefore risk assessment reports should avoid making statements on this question.

CONCLUDING REMARKS

The major findings in this chapter can be summarized as follows:

(*a*) There is no unique concept of risk that is used throughout all the risk assessment reports. Many of the important differences between the reports stem from alternative risk concepts used. Some reports do not even define their underlying risk concepts.

(*b*) The possible failures of a system, the probability of those failures, and the estimation of their consequences differ between the reports. Not all the differences can be explained in terms of the terminals and sites; some must be viewed as resulting from the limited knowledge and understanding of LEG risks. In this respect, most reports make scant reference to the remaining uncertainties in the estimation of risk.

(*c*) Although there are differences, there is no tendency for any single report to over- or underestimate the risk. Rather, each report is more conservative on certain topics and less so on others, compared to the other studies.

(*d*) On a relative risk scale it can be said that among the four sites Point Conception presents the lowest societal risk (because of the very low population density), Mossmorran—Braefoot Bay and Wilhelmshaven present the highest relative risk (because of the high population density and increased vessel traffic), and Eemshaven is in between.

(*e*) Although risk is an important aspect of the decision to import LEG and to choose a terminal site, other aspects such as reliability are also important. Any decision regarding LEG imports and terminal siting should involve comparisons with alternative options. As part of that process the risk of LEG should be compared with the risk of other options.

(*f*) Whatever the flaws of these LEG risk assessments, they are clearly superior to less systematic ways of identifying possible system weaknesses in order to inform decision makers on the topic of risk.

NOTE

1. The risk of multiple fatalities is typically displayed as a complementary cumulative probability distribution — the probability per year that the number of fatalities will exceed x shown against x. Such a curve, sometimes called a Rasmussen curve (see NRC 1975), contains information not available in the individual probabilities: the effect of correlations between those probabilities. A Rasmussen curve addresses the sensitivity to catastrophe found in the political perspective of risk. For instance, consider two facilities that can cause equal numbers of expected fatalities per year. In one facility these are bunched into very rare catastrophes, and in the other they are spread over common small accidents. The former facility may encounter greater political opposition due to sensitivity to catastrophe.

Chapter 8

Risk Analysis in the Policy Process*

Technological risks are big business. Tuller (1978) estimates that in the US the total damage in 1974 caused by technological hazards was in the range of $98 to 180 billion. According to a study by Clark University Hazard Assessment Group and Decision Research (1982), 17—31% of mortality in the US can be attributed to technology. Not surprisingly, risk analyses of technological hazards are growing in popularity. For instance, the US National Research Council (1981), which produces around 250 reports a year, estimates that half of these reports deal with risk and that one in five is a fully fledged risk analysis.

In our four case studies, we find no fewer than 15 studies of the risks from the proposed LEG terminals. The most important of these analyses were reviewed in Chapter 7. One conclusion was that risk analysts tend to present an overconfident picture of the accuracy of their estimates by the way in which they choose the data, couch the assumptions, and present the results. A second conclusion was that risk analyses are very narrowly focused on one aspect of the siting decision. What was called for was a comprehensive decision analysis of the alternatives (including the alternative of not building a terminal) in the context of multiple parties with their varied concerns.

Whereas Chapter 7 presented the analyst's perspective on improving analyses, this chapter looks at analyses in their social and political context. Although policy makers can be identified, no single decision maker can claim a comprehensive understanding of the social and political processes of which risk is a part. After drawing some general comparisons among the siting procedures in the four countries, we examine the sequential and interactive nature of the siting process, asking why more comprehensive analyses such as those suggested in Chapter 7, were not pursued. We then

*This chapter was written by Joanne Linnerooth.

turn to the uses and the usefulness of quantitative risk analyses in the political decision process. Finally, we ask whether analysts can usefully contribute to resolving complex public policy issues such as the siting of LEG terminals.

THINGS IN COMMON AND THINGS THAT GET DONE DIFFERENTLY

Drawing comparisons between the siting procedures in four countries, each with its own cultural style, governmental structures, and economic conditions is doomed to be both oversimplified and incomplete. To appreciate the difficulties involved, one might begin by comparing the observations of two participants in the IIASA Task Force Meeting of September 1980. William Ahern, a member of the California Coastal Commission, had this to say:

> I sense a sort of late 20th century technology. . . . The issues, the people involved, the kinds of things we are talking about, even the analyses seem to have much in common. Our differences are much greater on what we eat for breakfast or how we educate our children than on how we deal with liquefied gas. . . . (KLS 1982, p550)

Whereas Michael Thompson (1980a), a cultural anthropologist and member of the IIASA risk group, observes that

> Even a cursory comparison of the way the same technological risks are handled reveals that things do get done differently in different cultures (p.iii)

Depending upon how one goes about parceling or slicing up the case studies, staking them against each other reveals that there are surprising commonalities, but also that things get done differently. Our intent in this section is to distill out the issue of risks to the population, as distinct (or not so distinct) from issues such as national energy policy or public participation in technological questions, and to trace this issue through the four separate political processes. Setting the stage in this manner, we can turn to the similarities and differences in the national styles of handling risks, in the various roles played by the actors, as well as in the outcomes with regard to the safety of the selected sites.

An overview of the risk issue

The following brief review reveals that concerns about the risk of a proposed LEG terminal to the safety of the population was an important, sometimes even an overriding, consideration in the political debates in each country.

The FRG. The safety risk issue at Wilhelmshaven was an important concern during the early stages of the siting process, yet it appears to have been a somewhat less controversial issue than in the other three countries.

The 1976 settlement contract, where the state (*Land*) and local authorities agreed in principle to DFTG's plans to site an LNG import terminal at Wilhelmshaven, was signed in the absence of any comprehensive analysis of the potential risks of the facility. A review of the arguments in these first two rounds of interaction, neither of which included objector groups, reveals an atmosphere of concern, but of technological optimism. The safety of the terminal was considered by the parties involved to depend primarily on engineering factors. The terminal *could* be designed to satisfy existing regulations and would thus ensure the safety of the population. This picture, however, changed during the third round of discussions when public interest groups became involved and when the federal shipping authority (WSB) became concerned that a "residual" risk, especially to the residents of Hooksiel, would remain even if all the appropriate measures to ensure public safety were taken. The debate at the federal level, involving a number of ministries, resolved that this residual risk to the local residents and users of the recreational area near Hooksiel was acceptable.

The Netherlands. Doubts about the safety of an LNG terminal in Rotterdam harbor on the part of local authorities effectively diverted the decision momentum away from Maasvlakte and towards Eemshaven. The acceptability of the risk from the terminal at Maasvlakte was an important argument made by nearly all the participants throughout the three rounds of interaction. It was an issue that was addressed at an early stage of the decision process when the Ministry of Social Affairs commissioned a government supported research institute, TNO, to carry out a quantitative risk assessment for Maasvlakte and later for Eemshaven (TNO 1976, 1978). Yet the outcome, as distinct from the procedures leading to this outcome, seems to have been more a result of political and economic considerations, namely, the industrial development of Eemshaven, than of concerns about the risks to the public.

The UK. The safety risk issue of the LEG export facilities at Mossmorran—Braefoot Bay played a large role in the public siting debate, mostly as a result of the efforts of a strong and vociferous interest group that vehemently opposed the project. To appreciate the importance of the risk issue in the siting process it is important to recognize that the debate reported in the case study was geared to granting *outline* planning permission, or permission in principle for gas processing, storage, and transhipment facilities on the proposed site. Only if this outline planning permission is granted are full design details presented and the statutory authorities make a full scrutiny of the installations. Thus, the safety issue is not fully settled at that point in time where outline planning approval is granted. The main forum for resolving this question of planning permission was, in this case, the public inquiry, though a new safety issue emerged after the inquiry with the introduction of information concerning radio sparks.

The US. The safety risks of the proposed terminals were a dominant concern throughout the California decision-making process. A site in the Port of Los Angeles harbor, though supported by the local authorities, was turned down by the federal government on the grounds of an unacceptable safety risk posed by an earthquake fault. In Oxnard, a very different situation arose. The federal government was prepared to approve the site on the ground that the terminal presented an acceptable safety risk to the public, whereas the local authorities were more hesitant as a result of strong pressure from a local action group whose arguments had in many ways been reinforced by an environmental impact statement (SES 1976) prepared for Oxnard City Council. This report presented several worst-case scenarios showing the possibility, without accompanying probabilities, of 70 000 fatalities.[1] This scenario became the basis of a major push on the part of environmentalist interests for legislation requiring remote siting of LNG terminals. Pressure from environmentalists, along with pressure from industry and unions to expedite licensing procedures, resulted in the passing in 1977 of the California LNG Terminal Siting Act. Yet, even at the remote site of Point Conception, there were concerns about the risks posed by an underlying earthquake fault. This issue further delayed the process, to the point when the utilities have decided to defer their present application, on the basis that California no longer needs LNG.

Differences and similarities

While the safety risks of the proposed terminals were a prominent concern in each country, the *styles* by which the four political systems dealt with this concern differed. Keeping in mind that the four countries belong to the capitalistic, Western world, we find nonetheless differences in the degree of govenment centralization — from the highly centralized system in the UK, the consensual system of the Netherlands, to the more decentralized, federal systems in the FRG and the US. In the US, there is a more adversarial style of resolving conflicts, in contrast with the UK and the Netherlands where more emphasis is placed on discussions between the conflicting parties. Another point of interest is the expanding role of the courts in settling technological conflicts, especially in the US and the FRG.

Notwithstanding the contrasting national styles, we find that the stands taken by the various parties and the roles they played in the policy procedures showed remarkable similarities.

(1) The utilities, though highly regulated in the US, less so in the FRG and the UK, and partly nationalized in the Netherlands, served as the

initiators in screening and choosing the site(s) first submitted to the approval process. *Except for general planners, there existed no regional or national planning authorities for designating appropriate sites specically for an LNG terminal.*

(2) With the possible exception of the Oxnard City Council (where the issue never came to a vote), *the local city councils were in favor of having a terminal in their community.* In contrast with many siting controversies over nuclear power stations, we find in each case study that local authority representatives argued that an LEG facility would increase business and bring jobs to the local community.

(3) *The opposition to the LEG facilities was dominated by the "not in my backyard" (nimby) groups.* The Oxnard Citizens' Group and the Bixby and Hollister Ranch Associations in the US; the Hooksiel Citizens' Group in the FRG; and the Aberdour and Dalgety Bay Joint Action Group in the UK, were the most important. With the exception of the Sierra Club in California and the Conservation Society in the UK, there were no national or regional environmentalist groups actively involved in the siting debates. Though disinterest on the part of broadly based groups can be partly attributed to their often limited resources, that in many cases had already been allocated to opposing nuclear power plants, another contributing factor may be that liquefied energy gas (LEG) does not raise, at least as intensely, the same moral issues as nuclear power. The scale of a possible catastrophe is perceived by the public as more serious in the case of a nuclear accident, which involves longer-term, even intergenerational consequences. In contrast with the nuclear power controversy, the problem raised by the case studies lies in explaining the limited opposition, rather than any widespread opposition.

(4) *National and regional authorities played an important role in deciding whether the proposed facilities posed an acceptable level of risk to the workers and the public.* In the US, the federal authorities decided the fate of Los Angeles harbor, the state legislature decided Oxnard's fate, and the state and federal authorities ruled on the seismic risk at Point Conception. In the UK, the acceptability of the risks posed by the planned facilities at Mossmorran—Braefoot Bay was decided ultimately by the Secretary of State for Scotland; in the FRG this issue was decided by the Federal Minister of Transportation; and in the Netherlands the location of the terminal with its inherent risks was ultimately resolved by the cabinet.

While the actors and the roles they played appear similar, things still got done differently in different countries. Consider, for instance, California, with a decade of political controversy and no approved site for its LNG terminal. While this case appears exceptional, it should be kept in mind that the California case presents not one, but two separate siting procedures. The first was an unsuccessful attempt by the utility to locate at

Los Angeles, Oxnard, or Point Conception; the second attempt to locate at Point Conception represented a distinctly different process set by the conditions of the California LNG Terminal Siting Act in which the local authorities relinquished power to the state in the form of the California Public Utilities Commission. Today the procedure in California is possibly the most expedient of the four since the utility need only make "one stop" at the state level and "one stop" at the federal level, thus bypassing local public bodies. Hence, a process that was considered cumbersome by industry and valuable by certain local interests has been radically changed by this Act in the interests of minimizing delays in siting LNG facilities.

What is safe enough?

As a final comparison, we can examine the *outcomes* of the siting debates. Why were terminals considered safe at Wilhelmshaven, Mossmorran—Braefoot Bay, and Eemshaven, yet not at Oxnard or Rotterdam? The explanation may lie in the existence (or nonexistence) of less risky or otherwise more desirable sites. For example, the Netherlands and the FRG are limited in the number of remote sites, though in the Postscript Michael Thompson points out that the population density of some parts of the Scottish coast is similar to that of the California coast.

Another explanation can be found in the *economics* of the proposed sites involved. The communities of Wilhelmshaven, Cowdenbeath (near Mossmorran), and Eemshaven were in serious need of economic investment, whereas the Rotterdam and Oxnard communities were more affluent and more ready to question technological developments. As pointed out above, the decision on the acceptability of the risks was to a large extent the responsibility of the regional or national authorities; yet these authorities needed the support of the local communities and thus needed an appropriate justification for their decision. For this purpose, we find that the prospect of *jobs* provided a persuasive political argument in those processes where the sites were found to be "acceptably safe".[2]

Viewed in this economic context, the outcomes of the debates might be interpreted as following a simple and predictable pattern. If the harbor and the surrounding region are badly in need of economic development, the jobs that will be created by the terminal become a more persuasive argument and the risks will be viewed as acceptable. If the surrounding area is affluent or less in need of economic development, and if there exist reasonable alternatives, the risks of the proposed terminal become more salient and are likely to be found unacceptable. There is thus no absolute level of acceptable risk; rather, there is a close relationship between the question of the acceptability of risks and the economic setting in which these risks occur.

This is not a surprising nor undesirable finding since all questions of safety and acceptable risk should, at least from the economist's view, be considered in relation to the benefits gained by those who must accept the risks. Yet, on closer look, these outcomes do not reflect a clear and simple economic trade-off, but point instead to an opaque and complex process of political bargaining. First, as was pointed out by Robert Vincent from Gaz de France, an LNG import terminal does not create a significant number of long-term jobs.

> [The] terminal located near St. Nazaire has been, from the beginning, wished by all the people in this area, including the public as well as the Administration. It was believed that such a facility could be of some help for the unemployment situation in this area. Actually, I don't think an energy terminal provides a lot of jobs, except during the building period. (KLS 1982, unpublished draft).

Secondly, the economic trade-offs involved in accepting a risky technology were not made explicit; rather, there was often a pretense of an absolute level of safety against which the acceptability of the site could be measured. This notion of "acceptable risk", however, can only be an outcome of a decision process since there is no "objective" means of determining what is or is not acceptable. As Volker Ronge (1980, p231) puts it:

> The "question" of safety is no question at all — and the same could be said for the "question" of acceptable or unacceptable risk — and cannot be "answered" in an ordinary sense. . . . Instead, safety — as well as admitted, tolerated accepted risk — is an *outcome* of [a] highly complex systems process.

In sum, we find on one level starkly contrasting elements of style — the more adversary environment in the US, the highly centralized system of policy making in the UK, the emphasis on the consensus in the pluralistic Dutch system, and the legalistic federal system in the FRG. On another level, we find many similarities in the proceedings, especially in the positions taken by the parties involved. The project is initiated by a utility, supported by the local municipality for reasons of increased economic activity and jobs, opposed by a local environmentalist or public interest group, and eventually approved or not approved by the regional or national authorities. Underlying this process there appears to be a common and fundamental theme, which states that the acceptability of the risks will depend to a large extent on the economic conditions of the local community.

However, the economic interpretation for the selection of the sites, that the risks to the local community are balanced in some manner with the benefits to the local community, must be made cautiously and with some important qualifications. Though the prospect of jobs and local economic development was a persuasive argument for locating the terminal at a chosen site, the argument was often deceptive, and the implied risk—benefit trade-off was never presented as such. To some extent this economic interpretation presupposes that the siting question is a "decision" to be resolved by a single person with a clear knowledge of the trade-offs involved. The question is not, however, resolved by a "decision maker" at a single point in time; rather, it is addressed in small, sequential steps by a number of competing interests, whose interactions are colored by the political and cultural setting in which they occur.

THE POLITICAL CONTEXT

Sequential decision making

The siting of technological facilities is a policy issue that is not resolved as a systematic problem involving a range of costs and benefits, but rather as a sequential set of questions, each addressing small segments of the problem. How these questions are formulated by the "governing body" of that organization with responsibility for setting the formal agenda determines the considerations and concerns brought into the debate and therefore the types of scientific analyses generated. For instance, if the problem at hand is formulated as a choice between site *X* and site *Y*, the content of the debate, including the analyses, will be more narrowly focused than if the problem is framed in more general terms of identifying an appropriate energy source or an appropriate site along the coastline. As Braybrooke (1974) points out, the latter formulation is more appropriate for economic thinking, and for broader "systems" analyses that look at the costs and benefits of various alternatives; yet, it demands more complicated solutions and is hence less expedient. The choice between two sites, on the other hand, can be approached with simpler, incremental heuristics and, therefore, can be supported with more narrowly focused analyses.

Looking over the "problem formulation" (at least the visible agenda, as opposed to the "hidden agendas" of the separate parties) of each political discussion round for each country, it is possible to distinguish the five categories listed in Table 8.1. The content of the debate in the respective rounds was generally addressed to one or some combination of these

Table 8.1.

Problem formulation	Question(s)
Policy	Would a site be desirable from the national perspective?
Screening	Which site(s) or category of site(s) is (are) appropriate?
Selection	Is site *X* preferred to site *Y*?
Approval	Should site *X* (or site *Y*) be approved?
Licensing	What changes are needed in the site or technology before construction or operations can proceed?

problem formulations, but varied considerably across the four countries. Discussions in the US and the Netherlands covered an agenda from energy policy in the early rounds through site approval in the later rounds, whereas rounds in the FRG and the UK covered a somewhat narrower agenda, including only site approval and licensing. The siting question in the US and the Netherlands was therefore debated, at least in the early rounds, in the context of *energy policy*. On the other hand, in the FRG and the UK, the siting question was debated primarily in the context of *regional development policy*. This does not mean that energy policy and site screening questions were not addressed in the FRG and the UK, but that they were not addressed in a political forum or in the context of LEG.[3]

In this larger context, the siting question takes on an added status. It becomes a matter of *public policy*, which, according to Majone (1984) can be differentiated from individual decision making in many ways, the most important being that competing institutions take stands on policy issues consistent with objectives related to the long-term survival of their institutions. While the problem may be formulated as approving a certain site, other institutional concerns related, for instance, to energy policy or regional development may determine a party's position on the narrower agenda item.

Not only are the larger problems — whether and where to site an LEG facility — broken down into smaller subproblems, but these subproblems are usually dealt with sequentially by agencies with different and sometimes conflicting responsibilities. Constraints due to legislative and legal considerations may dictate the order in which certain actions must be taken. These decisions or actions are usually hierarchical, much like the sequence shown in Table 8.1. Resolution of the question of whether an LEG terminal is needed usually precedes the site selection phase which, in turn, usually precedes the licensing phase. Because of time and cost considerations, a decision on one level is often binding in that it cannot easily be reopened for political discussion. Thus, the process becomes tied or locked in to certain courses of action.[4] The responsible agencies have little alternative but to consider increasingly narrow aspects of the problem.

This locking-in phenomenon was illustrated in the US case study, where the California LNG Terminal Siting Act legally committed the state to the need for an LNG facility and to finding a remote site. However, the deregulation of natural gas prices in 1978 led to an increase in the supply of domestic gas. Rather than re-examining this need, the authorities pursued the question of whether there was a serious earthquake fault underlying the site, and the utilities have now withdrawn their current application on the basis of more optimistic forecasts of domestic natural gas supply.

The framing of the subproblems, as well as the order in which they are considered, may affect the final outcome. The sequence of events in the Netherlands illustrates this point. In an early stage of the process, the City of Rotterdam authorities commissioned a study revealing that the import of LNG would be cost-competitive with other import possibilities (e.g., gas pipeline) if it were on a scale of 25 billion m^3 per year (Rotterdam 1977). The order of the decision process was such that, first, a commitment was made to build the facility and, second, its scale was decided. As it turned out, the scale of the facility was reduced to below the cost competitive level. Had the order of the decision been reversed, it is possible that the project would not have been commissioned.

As these examples illustrate, a sequential decision process precludes the use of a comprehensive analysis of the LEG question in which the costs and benefits of various options could be assessed. Such an analysis would address questions ranging from energy policy through licensing *simultaneously*. From a political standpoint it is difficult to identify a site without some commitment on the need for the terminal and, likewise, to gain sufficient information on the technical licensing requirements without at least a partial commitment on the site.[5] The sequential process contributed to the fact that the sites were not chosen on the basis of some explicit balancing of the full range of costs and benefits.[6]

Party interactions

Understanding *how* a site is settled upon rests not only in recognizing the sequential nature of the process, but also in understanding the ways in which the interested parties take stands on the issues as they arise. A critically important element in this process is the extent to which each party with standing or decision responsibility can make a credible and justifiable argument to support whatever stance it has chosen to take. Justification for an LEG project that is viewed by the government officials to be in the national interest appeared to rest throughout the case studies on two *separate* arguments. First, the project would benefit the national economy and create jobs for the local community; and, second, the project would not impose an unacceptable risk on the local population. These arguments were independent of each other in the sense that the risks were not argued to be acceptably low because the terminal would create jobs, but rather they were acceptable due to some other criterion. Risk—benefit comparisons did not appear as a legitimate justification for a party's stand on the safety of the terminal. The question of "acceptable" risk thus became an issue unto itself.

As our case studies have shown, the question of whether a proposed LEG facility is acceptably safe is by no means clear cut with well defined criteria for its resolution; rather, it is intertwined with other concerns and is decided by a complex, interactive process. During the course of this process, those parties included take stands or argue for or against the safety of the proposed operations. The sophistication of these arguments depends on the manner in which these questions are usually resolved in the country concerned; in our case studies, the arguments ranged from showing that the facility was safe in the best judgment of the engineers, to comparing quantitative estimates of the risk with estimates of other risks or with some acceptable risk criterion. We review these arguments below.

Best engineering judgment (the terminal is safe). Many of the party positions on the safety of the proposed terminals were *not* based on quantitative risk analyses. Statements to the effect that LEG could be transported and stored safely at the sites under consideration were typically made during the early rounds of discussions by an applicant/developer in seeking approval for its preferred site. Often these statements were based only on the judgment of the engineers employed by the utility or agency who were knowledgeable about the operation of the planned project. For instance, in the FRG, when requesting and receiving planning permission (the settlement contract), the DFTG argued that there was *no* danger to the public; in the UK Shell/Esso argued successfully at the public inquiry that the safety of the plant would be ensured. These statements were based upon "best engineering judgment", supported by the reputation and safety record of the applicant.[7]

The Cremer and Warner report, commissioned by the Scottish local authorities to assess the safety of the proposed Mossmorran—Braefoot Bay terminal, was carried out much in this spirit. Although safety risk was recognized as a probabilistic phenomenon, no attempt was made to calculate these figures in a quantitative manner. The chance of an event that might lead to serious consequences was expressed in terms of low, very low, or extremely low, and the consequences were not enumerated in terms of, for example, lives lost. Nevertheless, the report did lay out the possible events that could lead to a spill of LEG at Mossmorran—Braefoot Bay, and therefore provided some detailed information rather than unqualified statements of safety.[8]

As technical problem solving moves further into the political arena, the difficulties presented by relying on principles of "best engineering judgment" or "reasonable man standards" become apparent. Lave (1981) refers to this practice as "intellectually bankrupt", and Rowe (1980) calls it an "old boys" form of decision making. Critics of this approach point to formal analyses, such as those discussed in Chapter 7, as an alternative. In fact, a risk analysis is not so much an alternative to using the judgment of experts, as a method for making this judgment open and transparent. The experienced engineer, when asked to consider the safety of an operation, will rely on his or her judgment of the likelihood of an accident and the seriousness of the consequences based on his or her experience with similar technologies or components of the technology. Unlike formal risk analyses, however, the cognitive processes of the expert making these judgments are not accessible to those relying on them.

Risk judgments (the risks are low). Turning to those arguments that are based on a risk estimate, meaning a quantitative or qualitative estimate of the probability of an accident and its consequences, we find various ways of introducing value judgments into the presentation of these arguments. The most subtle involves the simple choice of words: statements such as "the risks are negligible" or "the risks are high" introduce values by the way in which these words are interpreted. For example, the WSB in the FRG found the risks to be "non-negligible", which meant a need for concern, whereas the SAI consultants in the US found the risks at Oxnard to be "extremely low", meaning no need for concern.

Wherever low-probability, high-consequence events are concerned, values are inevitably introduced into the presentation of risk figures by the emphasis placed on either the probability of the occurrence of an event or on its consequences. The Hooksiel Citizens' Group in the FRG argued that the risks of the planned Wilhelmshaven terminal would be unacceptably high due to a significant probability of shipping accidents, whereas the regional government, as well as the Brötz *Gutachten*, argued that there was no danger since the worst conceivable accident would have no serious consequences. The Cremer and Warner risk assessment commissioned by the local authorities in the UK considered the probability of destructive events only and did not assess the consequences of the events. In California, representatives in the state legislature, in support of remote siting, placed the emphasis on the consequences, arguing that a terminal could not be located in an urban area where the possibility existed, no matter how small, for a catastrophic accident. In general, we find that in the case of low-probability, high-consequence events, those opposed to a project tend to emphasize the potential catastrophic consequences, while those promoting the project tend to emphasize the very low probabilities.[9]

Risk comparisons (the risks are lower than . . .). Frequently in debates on novel technologies the estimated risks are compared with those from natural or other technological hazards. In the US risk analyses, for example, the reader finds the individual probability of death from the LNG terminal being compared with the probability of being struck by lightning or of being the victim of a tornado. Although comparisons of this sort only provide a perspective without explicitly judging the acceptability of the risk, the message is that if the public is living comfortably with the dangers from lightning, an LNG terminal might also be considered acceptable.

A form of this revealed-preference approach to the problem can be found in each of the LEG siting debates, but was most prominent in the Netherlands. In the TNO study the quantitative risks estimates for

Maasvlakte (and later for Eemshaven) were compared with other industrial risks in the area, and the report concluded that the LNG facility would not add significantly to the existing cumulative risk burden of the population. This conclusion was cited by both STUNET (North Sea Island and Terminal Steering Committee) and the City of Rotterdam authorities in their support of the terminal.[10]

It has become increasingly clear that this revealed-preference approach to the problem may be helpful, but is not by any means sufficient, for determining the acceptability of a large-scale technology. Many of the concerns germane to the debate cannot be addressed by comparing the quantified risks, in terms of the probability of an event and the associated deaths, alone. A risk cannot be fully described by the probability of death; rather, risk is a multidimensional concept involving social, psychological, and cultural aspects that may somehow relate to the future of a high-technology society. These issues will be discussed further in the next section.

Acceptable risk criteria (one in million is safe). A seemingly more straightforward way of justifying the acceptability or nonacceptability of a risk is to compare it with an acceptable risk criterion, recognizing that such a criterion is essentially arbitrary. A number frequently found in the literature (see, for instance, Keeney 1980) is an annual probability of death of one in a million, meaning that an annual probability of death to a person that is equivalent to or below 10^{-6} is acceptable, otherwise unacceptable. This convenient figure can be traced back to an early paper by Starr (1969) who showed that the risk to an individual from natural hazards ranges between 10^{-6} and 10^{-7} throughout the world; hence this measure could be taken as a *de minimus* level set by nature. The US Federal Drug Administration, for example, in adopting this figure as an acceptable lifetime probability of death to a person from a carcinogen, stated that it was conservative in the interests of protecting human health (see Vaupel 1981b). More recently, the US Nuclear Regulatory Commission has also proposed an acceptable risk level for nuclear power plants, which states that nuclear reactors are safe enough if they add no more than two chances in a thousand to the risk of dying from other causes (*New Scientist* 1982).

Such a criterion was suggested only once in the case studies. The Aberdour and Dalgety Bay Joint Action Group, which carried out a risk analysis showing the annual probability of death to those people living in close vicinity of the Braefoot Bay terminal to be 10^{-4}, suggested that an acceptable risk criterion be set by the public authorities. The group stated that it would consider 10^{-6} to be a useful and fair yardstick for this purpose.

In sum, the interested parties supported their stands on the safety of the proposed LEG terminals based upon the judgment of engineers or upon quantitative or qualitative estimates of the risks involved. In the latter case, the risks were presented as high or low, significant or negligible, or they were compared with those from other natural causes, with those of other technological hazards, or with some arbitrary acceptable risk criterion. One of the most remarkable findings of the case studies is that in no case were the risks of the proposed terminal explicitly compared with the benefits. Since the economist would argue that a risk—benefit comparison is the only relevant one from a social welfare perspective, it is thus notable that such a comparison did not constitute a legitimate policy argument.

This appears especially puzzling to those who have witnessed policy making in other areas involving risks, such as, for example, many decisions in the US on highway safety, where risk—benefit calculations have played an influential role. Yet, even in the US and the UK, where cost—benefit calculations are a more acceptable input into policy decisions than in the FRG or the Netherlands, we find that weighing the risks against the benefits is becoming an increasingly questionable analytical practice. Such a practice inevitably involves placing an implicit or explicit value on human life, a procedure fraught with difficulties (Linnerooth 1975, 1979, 1982). Rarely have reliable quantitative assessments of costs, risks, and benefits been available, and all too frequently these factors are separated in time, accrue to different population groups, and generally appear to be incommensurable.

We concluded in the last section that the sequential nature of the decisions to site an LEG terminal precluded the usefulness of comprehensive analyses of the costs and benefits of the various alternatives. A narrower or more targeted risk—benefit analysis, barring the analytical difficulties of valuing lives, etc., might have presented a useful political *argument* except for the uncomfortable fact that the risks would be borne by a small group of people living near the facility and the benefits reaped by a larger population. In contrast, for example, with highway safety measures, the siting problem is fundamentally a problem of *equity* for which risk—benefit tools are not helpful, at least in the absence of any mechanisms for redistributing the costs and benefits.

It is not surprising, therefore, that the interested parties did *not* cite the overall distribution of the risks and benefits as a political argument, but cited only those benefits that were of direct interest to them or to their constituency.[11] To the local politician the benefits of providing a secure gas supply to the region or nation generally took second place to the benefits of providing jobs and economic development to the local community. The analytical challenge of finding a meaningful way to address problems where the costs and benefits accrue to different groups of people is taken up in Chapter 9.

THE USES OF QUANTITATIVE RISK ANALYSES

The process of *risk assessment* is generally thought of as comprising two separate activities: risk estimation and risk evaluation (Otway *et al.* 1975, Lowrance 1976, Rowe 1977, Jennergren and Keeney 1979). *Risk estimation* is concerned with identifying the various possible negative consequences of a project or activity and assigning probabilities to those consequences. This activity is usually viewed as the scientific, fact-finding part of risk assessment. *Quantitative risk analyses* (QRAs), such as those discussed in Chapter 7, are meant to contribute to this activity, without bringing in value judgments on the part of the analyst concerning the acceptability of the estimated risks. *Risk evaluation*, which examines whether the risks are in some sense "acceptable", is usually viewed as the subjective, value-laden part of risk assessment. In the previous section we examined how the various parties evaluated the risks in justifying their respective stands on the proposed sites. In this section we will take a step back and examine the uses made of quantitative risk analyses in estimating these risks.

Table 8.2. Overview of the major risk analyses.

	US (Oxnard)			Netherlands
	SAI	SES	FPC	TNO
Prepared for	Applicant	City Council	Federal Commission	Ministry of Social Affairs
Prepared by	Private consultant	Private consultant	In-house	Government-sponsored research institute
Timing	Round A (policy and approval)	Round A (policy and approval)	Round A (policy and approval)	Round A (policy and screening)
Use	Support applicant stand at hearing	Inform City Council	Support staff stand at hearing	Inform cabinet members and other parties
Scope	Vessel, transfer, and storage	Vessel, transfer, and storage	Only ship accidents considered relevant	Vessel, transfer, and storage
Methodology and format	Event and fault tree; quantitative probabilities	Composite of several studies; worst-case scenarios	Fault tree; quantitative probabilities	Event tree; quantitative probabilities
Conclusion	. . . risks . . . are extremely low	It is not now possible to state that [it] poses a low probability of a high-consequence event.	. . . risks . . . are negligible . . . an acceptable risk to the public	Societal and individual risk is low compared to other man-made risks
Influence	FPC persuaded to approve	Increased opposition	FPC persuaded to approve	Cabinet persuaded that both sites present acceptable risks

The widespread use of QRAs began in the US and their popularity is spreading to many European countries. Mazur (1980) traces the roots back to the development of systems analytical techniques, especially operations research and cost–benefit analysis during and directly after World War II. In the US, the National Environmental Policy Act of 1969 required all federally funded projects to be justified by environmental impact statements in which the various benefits and costs should be set out, preferably in quantitative form. With the emergence of the nuclear power controversy, the nuclear industry produced the Rasmussen Report (NRC 1975), which emphasized, in the spirit of an earlier article by Starr (1969), that nuclear power presented very small – and acceptable – safety risks to the public. Given the lack of historical data on reactor accidents,[12] this study was based on hypothetical estimates in the form of fault tree or event tree analysis. This technique has served as the model for many subsequent risk analyses of the reliability of technologies for which no historical data exist.

Who commissions and who produces risk analyses? As can be seen in Table 8.2, which shows the major risk analyses that entered the siting debates in the four countries, the answer to this question is seen to

Table 8.2. (*continued*)

UK		FRG	
Cremer and Warner	Action Group	Brötz	Krappinger
Local authorities	Members of Action Group	Federal Shipping Board (WSB)	Federal shipping Board (WSB)
Private consultants	Private consultant	Certified expert	Certified expert
Round A (approval)	Round C (approval)	Round C (approval)	Round C (approval)
Evidence at public inquiry	Prepared too late for official use	Advise WSB	Advise WSB
Vessel, transfer, and storage	Primarily jetty operations	Vessel, transfer, and storage	Vessel, transfer, and storage
Qualitative probabilities; no estimation of fatalities	Historical data and quantitative probabilities	Qualitative probabilities; no estimation of fatalities	Historical data and computer model; quantitative probabilities
No reason to doubt that the installation can be built and operated in such a manner as to be acceptable in terms of community safety	Individual risk is high compared to other man-made risks	With regard to the consequences and their probabilities, there is no danger, bearing in mind the relevant laws	Probabilities of shipping accidents estimated
Secretary of State for Scotland convinced of safety	Reinforced Action Group's view that risks were unacceptable	Did not fully persuade the WSB of facility safety	Informed WSB of probability of shipping accidents – considered high

differ in each case study. In the UK and the US, local authorities and interest groups reach into their pockets to pay *private consultants* for expert advice. In stark contrast to the number of reports produced in the US and the UK, the Netherlands appears to show a greater trust in expert authority; only one major risk analysis was carried out for Maasvlakte and Eemshaven by a *government-supported research institute.* This report was read, and frequently referred to, by nearly all the parties in the Dutch siting process. Still another style of recruiting expert advice can be found in the FRG, where traditional trust in scientific expertise is illustrated by the practice of commissioning *certified experts* to estimate the risks.

At what stage in the siting procedures were these risk analyses commissioned? The timing of the analyses varied significantly in the four countries. Only in the Netherlands was a quantitative risk assessment carried out during the screening procedures before a definite site had been identified and submitted to the relevant authorities for approval. In contrast, in Scotland only a generic, qualitative report (Cremer and Warner 1977) was submitted as evidence at the public inquiry. As another example, the report prepared by the UK Action Group was not given publicly acknowledged scrutiny by the statutory authorities since it was prepared after the public inquiry and was thus rendered inappropriate by statutory procedures.

Why were the risk analyses produced in the US and the Netherlands more quantitative than those produced in the UK and the FRG? The qualitative approach taken by Cremer and Warner in the UK was explained above as resulting from the nonavailability of detailed plans of the facility. A second contributing factor is, no doubt, the changing *philosophy* in the UK. In the past, decisions on plant safety have been left to the engineers or technicians responsible. Only recently have the statutory authorities required detailed, quantitative evidence of the reliability of technological systems.

In the FRG, the job of the regulatory agencies is to ensure that facilities meet the legally required standards for fire resistance, construction reliability, and so on. Certified experts are often asked to prepare reports that address these legal questions. It was somewhat unusual for the federal WSB to become concerned that the facility not only complied with existing safety regulations, but also that it did not pose the potential for a large-scale accident, especially through the synergistic possibilities presented by neighboring industrial facilities, and that it did not present a serious threat to the lives of local residents. The WSB, however, was not interested in the probability of a local citizen dying *per se*, and thus requested that Brötz carry out a qualitative assessment of the likelihood of events (in terms of low, very low, etc.) that might lead to a catastrophic accident.

Several other factors contributed to the lack of a comprehensive, quantitative analysis of the risks of the LNG terminal at Wilhelmshaven. First, responsibility for public safety is spread over a number of federal, state, and local agencies. The WSB was concerned with the risks from LNG shipping operations in Jade Bay, and this part of the problem was extensively analyzed. In contrast, somewhat less attention was paid to the possibility of a storage tank rupture, as a result of which a vapor cloud could threaten the population living near the site. One reason for this was that the WSB had access to the TNO report prepared for the Maasvlakte terminal, which showed that the risks from the shipping operations in Rotterdam harbor would be significantly more serious than risks posed by the land-based storage tanks. This result appeared to be true, as well, for the FRG.

Secondly, in the atmosphere of strong enthusiasm on the part of Wilhelmshaven authorities for the economic development of the area, there was little motivation for the city or the state authorities to take an independent and critical look at the plans and prepare a comprehensive risk analysis. This can be compared to the more skeptical atmosphere surrounding the deliberation of a site at Maasvlakte, which was economically more secure, and more concerned about the potentially adverse environmental consequences. Finally, in contrast with Mossmorran–Braefoot Bay and Point Conception, there was no strong interest group in Wilhelmshaven with sufficient funds to carry out its own risk analysis.

For what purposes were the risk analyses used? In answering this question, unlike those asked above, we find a great deal of similarity in the four countries studied. Risk analyses were frequently commissioned to support a party's position on the safety of the terminal. In the US, for example, the utility commissioned the SAI report in support of its position to the FPC that the terminal could be safely operated at Oxnard. As a second example, in the UK, the Cremer and Warner report was submitted by the local authorities to the public inquiry as evidence that the facility would be acceptably safe.

Majone (1984) draws the distinction between prospective (or predecision) analysis and retrospective (or post-decision) analysis, emphasizing that policy makers need the latter just as much as the former. In our case studies we find that often analyses played a dual role: for example, the report prepared by SES for the Oxnard City Council was used first to *advise* the council members and eventually to *justify* their positions to their constituency. The same dual role was played by the TNO analysis in the Netherlands, and by the Brötz and Krappinger reports in the FRG.

Analyses are thus conducted for multiple purposes and for multiple audiences, but are almost always intended ultimately to persuade these who are responsible for setting the relevant policy. Majone (1978, p213) thus sees a role for the analyst as "a producer of policy arguments . . . more similar to a lawyer . . . than to a problem solver." Others, however, view the use of the analyst's expertise as a political weapon in the adversarial process as a misuse of this expertise (see, for example, Behn 1979).[13]

When risk analyses are commissioned for the purpose of strengthening and justifying the client's case (and this appears to be more the rule than the exception) it is hardly surprising that they are written in such a way as to be as persuasive as possible. From Chapter 7, we find that assumptions are hidden, the uncertainties are not calculated, the data are carefully chosen, and presentation formats are constructed to direct the reader's attention to one aspect or another of the safety of the operation. In this way, the values of the client or analyst may enter into the estimates of the risks, clouding the distinction between risk estimation and risk evaluation.

Who reviews the analyses? A disturbing finding of the case studies is that there did not exist a fully adequate review procedure in any country. This does not mean, however, that the analyses were not reviewed. In the US there was an important redundancy of effort that exposed many, but by no means all, of the shortcomings of the analyses. In the Netherlands, a form of review was ensured by the fact that the one major analysis was read by nearly all the interested parties, yet not all of these were qualified to review the report critically. In the UK, the opposition groups had access to the risk analysis commissioned by the local authorities before the public inquiry, yet it is difficult for such a group to hire expensive consultants to review such reports. However, objector groups did play an important, and sometimes critical, role in exposing possible weakness in the terminal and operations. In the FRG, the analyses were reviewed by the staff of the government agency that commissioned the reports, but since the reports were generally not distributed, the adequacy of this procedure depended heavily on the existence of a qualified staff. Moreover, the staff may have little motivation to assess critically the merits of outside expertise that has been commissioned for the purpose of supporting an agency policy. Procedural reforms aimed at ensuring that scientific analyses receive an adequate review are suggested in Chapter 9.

USEFULNESS OF QUANTITATIVE RISK ANALYSES

Now that we have examined the *uses* made of quantitative risk analyses, we can turn to examining their *usefulness.* Do risk analyses make things safer, make people feel that things are safer, make people feel confident that their institutions are providing for the public safety, or improve the quality of political debate? Alternatively, do risk analyses contribute little to improving engineering design, fog discussions on safety in mathematical complexity, provide only a pretense of factual knowledge, or simply address the wrong questions? As risk analyses increase in number, these questions are becoming more and more subject to debate (see, for instance, Conrad 1982). In what follows we review first the supporting case and then the case opposing the use of quantitative risk analyses, drawing upon experiences found in the case studies. We then draw some tentative conclusions in the final section regarding the usefulness of quantitative risk analyses for addressing the issues of the safety of large-scale technologies.

The supporting case

Many analysts have argued that risk analyses, if applied with common sense, a sound knowledge of the technology, and a recognition of the inherent uncertainties involved, can be valuable and useful tools (Farmer 1976, Ramsay 1981, Drake and Kalelkar 1981, Stoto 1982). An intelligent analysis may reveal inadequacies in the design of a technology, may make the decisions affecting public safety more open and the public officials more accessible, and may provide the basis for more informed debate.

The use of fault or event tree analyses requires a detailed listing of the possible accident chains. Even if probabilities are not assigned, this exercise may expose a non-intuitive chain of events that could lead to an accident. Whether or not this information changes the analyst's view or judgment of the overall safety of the technology, it may help to identify the kinds of failures most likely to occur and to indicate options for reducing the risks. Some design problems may then be eliminated early in the planning process.

This preventive value of a QRA was illustrated by the FRG shipping authority's (WSB) use of the TNO risk study (prepared in the Netherlands for the Maasvlakte terminal) in identifying shipping accidents as a particularly weak point in the reliability of the operation, and their eventual decision to change the configuration of the deep-water shipping channel. Getting the design right on the first try is clearly commendable, though Flint (1981) observes that relatively few reported failures are the result of oversights in design; rather, most are the result of human error.

A risk analysis may serve a different, more modest role in helping planners gain a general picture of the risks involved, as expressed by Robert Vincent, from Gaz de France:

> For our risk analysis, our engineers simply adapted figures from other analyses done for the oil industry, although these figures could not apply to the kind of equipment we use for LNG. But we did use the results of our risk analysis. A figure such as 10^{-15} means perhaps 10^{-10} or 10^{-18}, and it means that the risk is actually very remote and not to be compared to value of 10^{-5}. This seems to be a good way to detect the weakest points of the system under study. We use risk analysis rather for this purpose than as a means to demonstrate that there is no risk or that there is only remote risk. (KLS 1982, unpublished draft)

One of the important functions of a QRA is to provide written documentation of the range of considerations underlying safety judgments, including the assumptions, the available data, and the "best guesses" of the experts. A risk analysis thus provides information to those who have access to the report and may result in more accountable public decisions. Such information is called for, since judgments by experts are becoming increasingly regarded as an unsatisfactory way of deciding upon the acceptability of large-scale technologies. According to O'Riordan (1981):

> Risk assessment becomes not just a device for seeking to alter the balance of political power, but also a means for reforming the nature of governmental secrecy, and agency accountability. Neither of these two vital functions will be altered rapidly or easily in any political democracy for many traditions and cherished positions are at stake. But the battle is on and the way in which risk assessments are used especially by the ecocentrists, is already proving an important strategic weapon in this quest for change.

In the Scottish case a comprehensive risk analysis prepared by the local authorities early in the proceedings may have made the procedure appear more accountable to the Action Group. However, Macgill and Snowball (1982) speculate that such increased accountability may not have an effect on decision outcomes or on the safety of a given plant; rather, the public relations value of a risk analysis, if adequately undertaken, may be more relevant. The Health and Safety Executive is responding to observations of this sort by making available more information to the public, though commercial confidentiality may inhibit the release, in full, of some of the material (Barrell, in KLS 1982, p320).

Where there are varying opinons within the expert community concerning the reliability of safety systems, a QRA might shed some light on the source of the conflict. For example, a source of conflict revealed by the QRAs carried out in the countries under study concerned the maximum downwind distance a flammable vapor cloud could travel following an instantaneous spill of LNG. These estimates, which may or may not have been crucial in assessing the danger to neighboring population centers, ranged from 2.3 km (Brötz, FRG) to 27 km (FERC, US) under differing atmospheric conditions. Since the physical properties of vapor cloud dispersion are not well understood, these differences cannot be resolved with the present state of scientific knowledge.

Information of this sort can be used in a number of ways. Industry and regulatory agencies might initiate further experiments to clarify the unknowns, as did the US Department of Energy in funding experiments on large spills of LNG. In those controversies where there are active opposition groups, this type of information might promote a more constructive dialogue between the public and experts focusing on the assumptions and data. In some cases, opposition groups might contribute to understanding the safety of the technology by carrying out their own independent studies, such as was done by concerned residents in the vicinity of Point Conception (the Hollister and Bixby Ranch Associations), and at Aberdour and Dalgety Bay.

The opposing case

The case that risk analyses, or analyses in general, contribute to reducing political conflicts is by no means proven. Mulkay (1979) reports on several studies (e.g. Nelkin's study of the Cayuga Lake controversy, 1971, 1975) investigating the uses made of scientific knowledge in the course of political debate. The main conclusion of these studies is that scientific knowledge contributes little to reducing the conflicts, but rather it becomes a resource that can be used to promote political aims. As we have seen in the case studies, opposing parties in disputes involving technical issues can generally obtain analyses from reputable scientists to support their positions. By now it is clear why this is possible — an analysis can be swayed in many directions by the simple choice of assumptions and the wording of the results.

In comparing the risk analyses of the four countries (see Chapter 7), one is struck by the large discrepancies in the results. This is especially apparent in the case of Oxnard, where several QRAs were carried out for the same site (see Table 8.2). The SES report concluded that "it is not now possible to state that [the facility] poses a low probability of a high consequence event." This is a markedly different conclusion from those of both the SAI and FPC reports, which found the risks to be "extremely low" and "negligible" respectively. The SAI study estimated the risks in terms of the probability of an individual death to be in the range of 10^{-7} to 10^{-10}, compared with the SES report where the results ranged from 10^{-4} to 10^{-7}.

Critics of formal risk analyses are quick to point out that the results are not objective, that different analysts will inevitably produce different results. It is apparent from the nature of the problem that there are many competent and respectable ways of estimating risks. No one set of assumptions is best, no analysis can be complete, and no assessments are "free" of judgment. Analysis is a social process influenced by human feelings, values, and beliefs (Meltsner 1980, Mazur 1980).

Although several authors have discussed the possible "pitfalls" of analysis (see, for example, Quade 1975, Majone 1980), whereby values on the part of the analyst color his or her methodologies and results and whereby heuristics introduce biases into his or her work, Brian Wynne (1982, p127) suggests that these biases be recognized as part and parcel of science, and not lapses from rational scientific analysis.

> There is a pervasive myth about the nature of science which supports this false approach to the question of "analytic bias". The tendency in the literature is to regard bias or mistakes as individual and isolated in origin, which suggests that ideal objective scientific knowledge can be attained in professional practice and as an

> input to policy issues .... This gives a fundamentally misleading and politically damaging picture of the role of expertise ....

The myth of scientific objectivity, especially where the policy sciences are concerned, has led to a dual perspective on risk analyses. On the one hand, because they are quantitative they appear to be factual or objective. On the other hand, the large uncertainties involved necessarily push the evidence out of the realm of facts and into the realm of what Weinberg (1972) has called "trans-science". This *dual nature* of a formal risk study has fogged discussions of its role in the policy process. The numbers produced by a risk analysis are not exact or "hard". They incorporate a number of judgments, but as has been pointed out by Ravetz, hardly anyone in our culture is capable of handling inexact quantities or "soft" numbers (see KLS 1982, p402).

Recognizing that the estimates produced by risk analyses cannot be fully objective, critics of risk analyses object further that the probabilistic results themselves are open to a number of sometimes contradictory interpretations (see Vaupel 1982a). It was noted above that in the case of low-probability, high-consequence events, objectors to a project frequently highlight the consequences and promoters highlight the probabilities. In no controversy was this more clearly seen than in Oxnard. The SAI report (commissioned by the utility) reported the results in terms of the probabilities of individual fatality or catastrophic consequences and compared these probabilities with those from other man-made or natural disasters. Since these probabilities were generally higher than those calculated for nuclear power plants (NRC 1975) the authors of the report were careful not to include this comparison.[14] Alternatively, the SES report (commissioned by Oxnard City Council) presented vapor cloud/population risk scenarios, which graphically illustrated a deadly methane cloud covering parts of Oxnard, with no accompanying probabilities. Whereas the SAI report persuaded the federal authorities to approve the Oxnard site, the SES report had the opposite effect, and was influential in persuading the state legislators to rule out Oxnard as a possible site through the remote-siting provision of the LNG Terminal Siting Act.

In the Netherlands, we find an example of a single report taking on different meanings according to the purpose for which it was used. The TNO risk analysis supported the arguments of those promoting the Maasvlakte site, namely, that the terminal would not add significantly to the cumulative risk burden of the local residents. The analysis was also used to support the case made by the Rijnmond Public Authority, that the proposed terminal *would* increase the safety risks to the local population. Most of the discussions by the local authorities focused not on the likelihood of an accident, but on its possible consequences, the worst of which could be over 17 000 deaths in the Rotterdam area. Because the TNO report contained evidence to support the case of the proposers as well as those questioning the project, there was little need for any party to prepare a counter-expertise.

This ambiguity in the results of risk analyses is also observed by O'Riordan (1981, p160) who notes "the curious spectacle of not only the 'doomsday' extremists but also the 'hard-economic facts' school of developers both using the same objective risk analysis data to prove their cases . . ." To some extent, all scientific knowledge takes on a meaning relevant to its social context (Conrad 1982), but Johnston (1980) argues further that risk research is an immature science, or one in which there is little consensus over fruitful lines of inquiry or even competent research. When an immature science is relied upon to solve a practical problem, there is a tendency for the outcome to establish the conclusion that its sponsors wanted. Yet, according to Jerry Ravetz, the science of risk analysis is maturing:

> Of course, in the earlier period when the theme was being developed in a totally undisciplined way without a lot of collegial criticism, people simply got some accident statistics together and put exponents on numbers. Then it was a game that any one could play, and any number could have been used . . . . I think today . . . we don't play that game any more . . . . It is now used as part of a negotiation rather than as part of propaganda campaigns. You can still disagree. You can still use it to mystify if you can get away with it. But it is no longer simply a case of valium from him and amphetamine from him. The thing is maturer; it has changed character. (KLS 1982, p404)

Ravetz is referring here to what Reijnders (1982) has termed "valium" and "amphetamine" analyses to reflect the resourcefulness of the analyst in painting radically opposing pictures with the available data. This is nicely illustrated with an example by Vaupel *et al.* (1982, p5), which shows that it is possible not only to *lie* with statistics but also to *insinuate* with statistics:

> The highest estimate of the probability of a major accident at the proposed LNG site at Point Conception, California, was that there was one chance in 10^{-4} that 10 or more people would die. Someone who wanted to make this risk look small might claim that the risk is the same as the risk of one death every thousand years. On the other hand, someone who wanted to heighten the risk might proceed as follows. The analyst who made the estimate of 1 chance in 10 000 per year of 10 deaths or more was unsure about this estimate — he thought that there was about 1 chance in 100 that the true value could be as great as 1 chance in 100. In ten years a risk of 1 percent per year accumulates to about 10 percent, and 10 percent might be called a "significant chance". The loss of 10 deaths or more was defined in the risk analysis report as a "disaster", so the headline could read: "Analyst says that it is possible that there is a significant chance of a disaster at Point Conception in the next decade".

Indeed, insinuation is nearly impossible to avoid because nearly all policy-relevant statistics take on somewhat different value colorings and import various policy thrusts depending on how they are expressed and how they are placed in context (Vaupel 1982a).

Despite the open-ended possibilities for interpretation, risk analysis often has the appeal of presenting a rational approach to public decision making (Moss and Lubin 1980). The precise numerical results may provide comfort by concealing the inherent and fundamental uncertainties, such as those reflected in the millionfold difference between risk estimates for saccharin and the Inhaber—Holdren debate (Inhaber 1979, Holdren *et al.* 1979) concerning the risks of nuclear power plants. This concern was expressed recently in a report surveying the way in which risk analyses are prepared by the US National Research Council:

> Science is strongly biased towards numbers, for when numbers can be justly employed they denote authority and a precise understanding of relationships. Because this is so, there is an equally important responsibility not to use numbers, which convey the impression of precision, when the understanding of relationships is indeed less secure. Thus while quantitative risk assessment facilitates comparison, such comparison may be illusory or misleading if the use of precise numbers is unjustified. (NRC Governing Board Committee on the Assessment of Risk 1981, p15)

As was pointed out in Chapter 7, LEG risk analyses are not immune to this criticism. For example, the probability of a spill resulting from a vessel accident at Point Conception was estimated by SAI as 9.9×10^{-7} and by ADL as 8.1×10^{-3}, without any accompanying discussion on the part of either report on the confidence intervals or uncertainties involved. Another telling example is found in the SAI report prepared for the utility, which reported the probability of a catastrophic accident at Oxnard to be 10^{-57}, or one chance in 710 septendecillion of a maximum catastrophe of 113 000 fatalities! It is difficult to imagine such a low probability, especially as it was expressed as a point estimate with no mention of its confidence interval. Yet this figure was quoted by the Federal Power Commission (FPC 1976) in justfying its support of the Oxnard site.[15]

A danger of client-oriented research is that those commissioning the analyses may be reluctant to fund research that is likely to uncover new uncertainties that they feel will only further complicate the decisions at hand. The clients may search for and reward experts who are willing to present precise and unqualified estimates of the risks, where the case is fairly secure. Or they may encourage the analysts to adopt the most conservative or worst-case assumptions, a practice that has been criticized by Raiffa (1980). As Stoto (1982) points out, analysts have an easier time justifying results that their clients want. Presenting the results in an overconfident way is particularly worrying since the uncertainties involved may be of more concern than the exact estimates of the risks. In Holling's view (1981b, p1), "The real questions are how to regulate and predict and how to live within the limits that our ignorance places on each."

Possibly the most damaging critique leveled at those carrying out risk analyses is that the problem of reaching a consensus between conflicting interests in society, e.g., over the location of a large-scale and novel technology, cannot be solved, or even helped, by expert calculations of the safety risks. The analyses are simply addressing the wrong problem. By defining risk as some combination of probability and physical consequences (usually lives lost), other social concerns about the technology, for which risk may only be a surrogate, are ignored. In other words, risk may only be a symbol for the fears and anxieties that the public may have concerning the technology. These fears are not adequately captured by the "body count" figures found in risk analyses.

The psychologist views these fears in terms of the perceptions of the public regarding the seriousness of a risk. Much research suggests that people are sensitive to many dimensions or characteristics of a risky activity as well as the probability of death (see especially Lichtenstein *et al.* 1978, Slovic *et al.* 1979, 1983, Fischoff *et al.* 1981a, Otway and von Winterfeldt 1981, Stallen 1981, Clark University Hazard Assessment Group and Decision Research 1982, Humphreys 1982). For example, whether the risk is voluntary or involuntary, whether there are potentially catastrophic consequences, whether the person has control over the outcome, whether the risk is observable, novel, equitably distributed, and so forth, are all relevant to how serious an individual perceives the risk to be. Stallen and Tomas (1981, p39) conclude that "there is no (psycho-)logic according to which people's assessment of technological safety is proportionally related to (or can be predicted from) observed relative frequencies or statistically calculated probabilities of negative consequences of the technology concerned."

Alternatively, the sociologist does not see the problem of personal risk assessment as depending so crucially on the perceptions of the seriousness of risk, but rather on social and political values, including plans for the future of society, the evaluation of political decision-making processes, the credibility of institutions, and communication of information (Nowotny 1982). According to Otway and von Winterfeldt (1981) risk has become so topical in today's debates about the acceptability of technologies that the complex problem of social acceptability is too often reduced to a mathematical problem of defining and measuring risk.

If opposition to LEG terminals is grounded in these vague misgivings about the future of technological society, the decision procedures in the four case studies discourage debate on this basis. The reason lies in the necessity, which is true to varying extents in the countries studied, to justify arguments on scientific (often interpreted as quantitative) grounds. A review of the arguments brought to the debate show that the majority, including the need for natural gas, the jobs created, and the economic benefits, are in some sense quantifiable.

Still another school of thought, developed by the anthropologists, explains problems of technology and technological risk as embedded in the different cultural biases (or political cultures) that make up society. Douglas (1972, 1978a), for instance, sees pollution or risks as threatening to some because the pollution or risk intrudes upon their standard of "purity". Rules against pollution or risk can only be understood as a defense of a specific "moral" order, or some state of society that is deemed worth preserving. Thompson (1980a, b, 1981) has developed these ideas into a cultural theory of risk, which he explains in the Postscript to this book. According to Thompson, the individual is not an isolated entity, but a social being, and how she or he interprets and behaves toward technological risks is shaped by this social context. He develops five cultural categories depending on the extent to which a person is bounded by social groups and the extent to which he or she is subject to socially imposed prescriptions. Although Stephen Cotgrove (1981) speaks of only two alternative paradigms to which an individual adheres, his interpretation of technological conflicts fits nicely with that of Thompson. Interpreting the nuclear debate, Cotgrove remarks:

> What is rational and reasonable from one perspective is irrational from another. If the goal is maximizing output, then nuclear risks are not only justified but it would be unreasonable not to take them. From another perspective, from the viewpoint of a quite different set of beliefs about how the world works, and quite different aims for some kind of more convivial society, to take even the possibly small, but in practice incalculable, risks for future generations stimulates a moral indignation which justifies unorthodox political action that crosses the threshold of legality. (p128)

It is this second paradigm that Ronge (1982) sees as threatening the very core of society in the FRG, as well as other countries that are experiencing a social movement referred to as "alternative culture". In this environment, Ronge doubts whether risk analyses commissioned by traditional political institutions can be helpful. "Risk researchers are players — and victims — of a game which is fundamentally opposed by the new social movements".

A remarkable aspect of the LEG debates recorded in the case studies is the conspicuous absence of this "alternative culture", especially the cultural category that Thompson refers to as the *sectist*, who is strongly bounded to social groups but relatively free from socially imposed prescriptions. The absence of sectist groups, such as the Friends of the Earth[16] can in part be explained by the type of risk presented by LEG. There is evidence in the case studies that LEG risks are not perceived to be as serious as those from nuclear power. Indeed, in the Netherlands, one argument that proved persuasive in dampening environmentalist objections to a Maasvlakte site was that if an LNG terminal was approved, this would preclude any possible future siting of a nuclear power plant in the area. Individual risks from an LNG terminal as estimated by the TNO report were, however, higher than the individual risks from nuclear power estimated by the Rasmussen Report (NRC 1975). If LEG does not arouse the same moral indignation as nuclear power, this explains to some extent why the LEG controversies remained relatively low-key.

In sum, the psychologist, among others, sees individuals reacting to risks along a multidimensional scale of the characteristics of that risk, whereas the sociologist and anthropologist, again among others, see risk in its broader political and cultural context. Jungermann *et al.* (1982) describe the development of two "camps" of disciplines engaged in risk research. One camp views the technical issues of risk analysis as side issues that cover up value and cultural differences underlying technological debates. The other camp agrees that this is true for some technological issues, especially nuclear power, but that there exist smaller-scale problems for which the technical questions are more relevant. The first camp is skeptical of analytical methods that may circumvent the political process, whereas the second camp believes that analysis is merely an aid to these political processes and that analysis can enhance communication between conflicting groups and increase the transparency and accountability of political decision processes.[17]

CONCLUDING REMARKS

Notwithstanding the merits of detailed, quantitative risk analyses for improving the design of a hazardous technology, the usefulness of these analyses for resolving the questions of whether or where to put the technology is somewhat less clear. The evidence found in the case studies suggests that technical risk analyses contribute little to building a consensus concerning the safety of a proposed LEG terminal. The results of these studies, though generally presented in an overconfident manner, were nonetheless susceptible to conflicting interpretations. There was thus a tendency for an analysis (or analyses) to polarize further the political controversy.

In response to the apparent deficiencies in the "science" of risk analysis, a bill (HP8303, the Risk Analysis Research and Demonstration Bill) has been introduced into the US Congress to establish a program for improving and facilitating the use of risk analysis. Where science cannot provide answers, the only recourse, as evidenced by this proposed legislation, appears to be to improve the science. But a risk analysis, no matter how sophisticated the methodology becomes, cannot provide unambiguous estimates of the risks involved; the numbers will remain speculative. Nor will such analyses be able to shed light on the value-laden question of whether the risks involved are acceptable. According to Moss and Lubin (1980, p29):

> We can't turn back the clock of political mood. We can't tell legislative bodies that they will have to wait for better science, or better risk assessment methodology, or better risk-balancing institutions (mechanisms) to develop from the better science.

On the other hand, we can't pretend that we are dealing with a fully developed technology, ready to be plugged into whatever problem society comes up with.

If the answer does not lie in improving the science, it may lie in improving the procedures by which these decisions are made. We must carefully examine the use of scientific expertise, with its inherent uncertainties, in policy processes that are ill equipped to cope with these uncertainties. Without discrediting the analyst's role as a policy advocate, we must ask what changes are needed in our institutions to encourage a more honest reporting of the limits of analytical expertise.

We must also recognize that conflicts over the future of technological society cannot be resolved by the scientists alone. Yet, there is hope that analysts can contribute usefully to resolving complex public policy issues such as LEG siting decisions. In addition to providing the necessary technical expertise, a new and promising role for the analyst lies in the area of conflict mediation. This, and other prescriptive measures for improving the siting process will be taken up in Chapter 9.

NOTES

1. An accompanying chapter of the SES report did present the risks of the terminal in terms of individual probability of death.
2. The Los Angeles harbor was also economically depressed but was found to be too risky by the *national* government. Los Angeles City Council, however, voted in favor of the terminal.
3. The rounds recorded in the case studies generally commence at that point in time when the question of siting an LEG terminal gains status as an issue on the political agenda. The term *agenda* here can be thought of as a set of questions or controversies that are viewed as falling within the range of concerns meriting the attention of the political authorities (see Cobb and Elder 1972). In both the FRG and the UK, the site screening and site selection stages were primarily of internal concern to the utility (Ruhrgas—Gelsenberg and Shell/Esso). Shell/Esso consulted informally with local, regional, and even national authorities before lodging a planning application, but, in the UK "the onus of site selection rests firmly with the applicant." (see Chapter 5). The energy policy questions in the FRG and the UK were, for the most part, considered resolved by all parties to the debate; the need to import LNG to the FRG and the desirability of exploiting North Sea oil in the UK were settled in rounds of discussions preceding the LEG siting issue.
4. For example, in the US, the applicant originally stressed the risk of an interruption in the supply of natural gas as a major reason for importing LNG to three separate sites. During the course of the decision process, the three sites were reduced to one, and the number

of storage tanks at that site were reduced from four to two. Because of this concentration in one small area, and the possibility of routine closures or nondelivery due to bad weather conditions, the net result of the sequential decision process was that a project originally meant to decrease that risk was shaped over time into a project that may increase supply interruption risk.

5. In the US the policy question to import LNG was resolved by the state legislature independently of whether an acceptable site could be identified. This was also the case in the Netherlands, where an official government policy paper on the need to import LNG preceded the site negotiations.
6. Of course, with more national or regional planning some of the sequential decisions listed could be exogenous to the question of siting a particular plant. As Norbert Dall from the Sierra Club points out, more state planning (as envisaged by the California Coastal Plans and San Francisco Bay Area industrial siting plan) may be a practical alternative to difficult case-by-case decision making (1982, personal communication). However, an IIASA case study carried out for the selection of a gas pipeline route in the USSR (Mechitov 1982) shows that a similar sequential apparatus exists in this centrally planned economy.
7. In neither the FRG nor the UK, however, was this approval final; rather, it was conditional on the eventual scrutiny by national authorities — the WSB in the FRG and the Health and Safety Executive in the UK.
8. As noted above, the lack of rigor on the part of the analysts can be attributed to the fact that they did not have access to detailed plans that would be made available at a later stage along with a complete safety audit.
9. Mary Douglas (1982, personal communication) has suggested that a third characteristic of debates of this sort is the tendency for the opposition to focus on the adequacy of the decision procedures.
10. The cabinet decision in favor of Eemshaven, however, did not make any statements on the comparability of the risk in support is its argument that the risks at Eemshaven were acceptable. In fact, the cabinet did not justify its argument of "acceptable risk" in an explicit manner.
11. O'Riordan (1981) argues that much of environmental impact analysis (EIA) in the context of energy developments has become institutionalized, even ritualized, without fundamentally altering the decisions of energy policy makers. He suggests, further, that one motivation or the promotion of QRAs over EIAs was that the dangers to human health were viewed as a more powerful argument than the dangers to the natural environment.
12. Even when limited experience does exist, Fairley (1981) has pointed

out that data of zero occurrences of a catastrophe is of cold comfort in ruling out a catastrophic event.

13. Vaupel (1981), in discussing procedures for setting environmental standards, draws a crucial distinction between the risk evaluation *processes* that produce health, safety, and environmental standards, and the risk evaluation *justifications* that are used to explain, defend, and advocate the resulting standards (p95).
14. Although the SAI report adopted the other comparison reported in the Rasmussen Report (NRC 1975).
15. An exception to this failure to address uncertainty is found in the SES risk analysis, which, by referring to the range of estimates in other reports, explicitly dealt with the inherent uncertainties.
16. The Friends of the Earth were not, however, absent from other controversies involving LEG. The controversial Staten Island terminal was the basis for some activity. See, for example, the highly critical book on this subect by R.N. Davis (*Frozen Fire*, 1979) published by the FoE.
17. Despite these differences, a broad-ranging research program including researchers from these two perspectives has been undertaken at the International Institute for Environment and Society, Science Center, Berlin.

Chapter 9

Improving the Siting Process*

The descriptive material presented on the four case studies illustrates how different interested parties form strategies and present arguments to defend their positions regarding the siting of technological facilites. This chapter has a prescriptive flavor by focusing on ways to improve both the decision process and the resulting outcomes.

In Chapter 1 we noted that there are two broad objectives that guide final choices: the *welfare* and *distributional* objectives. Each party will have a different view of the relative importance of these two objectives because each sees the problem from its own special vantage point. Potential conflicts emerge for this reason; the institutional procedures in each country determine if and how these differences are settled for any given problem.

To illustrate why these conflicts are likely to emerge, first consider the welfare objective. The applicant may argue that a new LEG facility can be justified from the point of view of societal well-being, in that it promises to spur regional development in an area, and will increase the security of future energy supplies in a cost-effective manner when compared to other options. The applicant's criteria for justifying the final decision are likely to be primarily economic, with the provision that the facility meets specified environmental and safety standards. On the other hand, public interest groups like the Sierra Club may argue that net benefits to society are primarily based on the impact that the facility is likely to have on the quality of the environment for future generations, and so they may reach the opposite conclusion regarding the desirability of the facility.

The distributional objective is likely to cause similar problems. Some local residents may oppose a project because they fear the consequences of an accident even if they agree that society will be better off having the terminal. They might favor the project as long as it is "not in my backyard". On the other hand, local government may want the facility because

*This chapter was written by Howard Kunreuther.

they believe that it will have a positive effect on economic activity in their area. Whether in fact it will achieve this objective is an open question, as was illustrated in Chapter 8.

Reconciling these differences may not be an easy matter. Louis Clarenburg, from the Rijnmond Public Authority in the Netherlands, pointed out the difficulties policy makers face in drawing up explicit rules for dealing with welfare and distributional concerns:

> How can you make up your equity equation? How can you compare apples and guns in a final analysis, and come up with a computer solution? For political decisions, I cannot use the computer at all because it is really weighing values, benefits, costs, who is suffering, who is gaining. I feel no computer can give this answer, and if it can it is gambling with our democracy. (KLS 1982, p370)

Although a computer cannot solve the equity problem, nonetheless decision makers can be assisted in dealing with the conflicts inherent in siting hazardous facilities by the use of policy analysis, which encompasses both risk analysis and other approaches for improving the siting process and the final outcomes. Its actual use will be determined in part by the nature of the existing decision process. Because of differences in the entitlements, standing, and responsibilities of interested parties within each country[1] there is no *one* right way to do business in the world. National styles are important when proposing policy remedies.[2]

A COMPARISON OF INSTITUTIONS

The institutional arrangements governing siting decision processes affect the types of analysis that will be used. Two types of process models can be contrasted:

- A *judicial model* involves a single decision maker in the form of a judge, commission, or government body who decides on the outcome after hearing statements and evidence from the various parties that have been given standing.
- A *compromise model* involves direct interaction between the different stakeholders involved in bargaining, and negotiation. The implicit rule for reaching a final decision is unanimity. In practice this is rarely obtained, although tools such as compensation can transfer some of the gains from potential "winners" to "losers". The four case studies reflect different variations on these two broad approaches to resolving siting conflicts.[3]

The FRG. In the FRG a proposed LNG terminal is licensed like any other major industrial project. The applicant must develop a plan for a particular site, which is then presented to the relevant authority (usually the *Land*, one of the ten autonomous states of the FRG). In the case of the Wilhelmshaven facility, Lower Saxony played a predominant role since it was not only involved in the licensing procedures, but was also directly concerned with the impact of the project on regional development. In addition, all facilities must comply with town and country planning regulations of the municipalities responsible for approval of the plans (in this case the city council of Wilhelmshaven).

Any individual or institution has standing by being able to appeal against a particular decision directly to the courts, which can rule on its legitimacy. In the case of Wilhelmshaven, the threat of a court case by the citizens of Hooksiel led to negotiations, resulting in a compensation agreement in which Lower Saxony increased its subsidy for a planned recreational center. There was thus an attempt to reach a compromise between the relevant parties even though the decision structure is characterized by a judicial model.

The Netherlands. The siting procedure in the Netherlands is centralized only to the extent that the national government coordinates major decisions. The existing institutional arrangement promotes a mixture of adversarial debates and consensual decision making. A large number of interested parties had standing and were involved in the LNG debate so that points of conflict surfaced and were aired; in this sense the Dutch siting process is in the spirit of a compromise model. At the same time the desire for consensus was expressed formally by the central position assigned to an interdepartmental coordinating committee (ICONA), which included representatives of all relevant ministries.

Since Gasunie is a semi-state-owned company it was operating in line with official Dutch energy policy when it applied for a site for an LNG terminal. Formal responsibility for site approval for industrial facilities normally rests with the provincial and municipal authorities. Opposition to the proposed sites at Maasvlakte came primarily from the Rijnmond Public Authority, a collection of 16 municipalities in the Rotterdam harbor area that was primarily concerned with the safety risks associated with LNG. The local authorities in Groningen, on the other hand, welcomed the prospect of a terminal at Eemshaven because it promised to stimulate industrial activity and provide jobs in this area. Environmentalists opposed the construction of LNG facilities in both areas, but these groups were unsuccessful in acquiring an influential role in the process. No citizens' action organizations were formed to protest the applications, although some of the residents in the affected communities linked up with the environmentalist groups. The Dutch cabinet, which had final authority on this siting question, selected the Eemshaven site knowing there was united official support for the subsequent approval of the terminal at provincial and municipal levels. In this sense, this procedure was more in keeping with a judicial model.

The UK. The UK has a well defined hierarchical structure for reaching decisions. A developer wishing to build a facility normally files a planning application at district council level and this proposal is then transmitted to the regional or national level whenever the project is expected to impact a wide area. The proposed LEG facility at Mossmorran—Braefoot Bay was considered to have national significance for energy policy and industrial development so that the relevant decision maker was the Secretary of State for Scotland. The decision process was thus in the spirit of a judicial model rather than one involving compromise. The district council played a coordinating role by publicizing the application so that interested parties could state their positions in writing prior to any official decision. Due to substantial opposition to the Shell and Esso applications, the Secretary of State decided to hold a public inquiry that would focus on local issues. The local authorities unanimously favored the application because of the potential positive impact on employment in the area. Following the public inquiry the Secretary of State approved the project despite strong objections by the Conservation Society and an action group composed of residents from villages bordering on the site.

The US. In recent years the decision process regarding the siting of hazardous facilities has been transformed from a technological regulatory process with cooperation between parties, into an adversarial system (Allison *et al.* 1981). This change has occurred because the courts have played a more active role so that all parties feel they have an opportunity to present their positions to an administrative law judge presiding over a set of hearings.

The California case study illustrates a change that was made to institutional arrangements for LNG siting decisions. When the Western LNG Terminal Company initially filed an application with the Federal Power Commission (FPC) in September 1974, it was required to obtain approval for any proposed site from a number of different sources: the FPC was responsible for ruling on the environmental and safety risks, the California Coastal Commission (CCC) was charged with preserving coastal resources, and a number of local agencies had the authority to issue permits. Due to an impending stalemate regarding approval at either Oxnard, Los Angeles, or Point Conception, the applicant pushed for new legislation. The result was California LNG Terminal Siting Act of 1977, which authorized one-stop licensing, so that Western was only required to obtain approval from the California Public Utilities Commission (CPUC) rather than from the CCC and local authorities. The Sierra Club, in conjunction with citizens' groups at Oxnard, strongly favored a remote site, preferably offshore, and their views were thus partially reflected in the legislation through new restrictions on maximum population density within certain prescribed limits from the terminal. These legislative changes reflected compromises between some of the relevant parties but moved the siting process into the judicial camp since the CPUC was the sole decision-maker.

THE ROLE OF POLICY ANALYSIS

We now turn to the question of how policy analysis can aid the siting process. Three perspectives are considered. From one perspective, facts can be provided on the potential effects associated with a proposed facility. From a second perspective, policy analysis can help elicit preferences of the parties with standing in a siting debate, so that decision makers may gain insights into what aspects are likely to produce conflicts. From a third perspective, policy analysis can help design proposals for clarifying the entitlements of the groups and the individuals affected by a siting decision. These options may help resolve some of the conflicts between the welfare and distributional objectives.

Perspective 1: Clarifying the effects

Analysis can help clarify the potential effects of a proposed facility in several ways (Vaupel 1982). At one level experts can provide an analysis of the costs and benefits of a particular technology such as the impact of importing LNG on the future price of energy, while at another level experts can provide data on the effects associated with locating a facility at a specific site. Examples of the latter include the consequences to the citizens of Mossmorran—Braefoot Bay in the event of a vapor cloud explosion following a storage tank rupture, or the costs associated with transporting LNG if Point Conception were chosen as a site in California. We have explored in detail the discrepancies between experts regarding risk analyses for LEG siting in Chapters 7 and 8. As was pointed out in Chapter 8, there did not exist a fully adequate review procedure in any country. In addition, objector groups played an important, and sometimes crucial role in exposing possible weaknesses in the safety of the terminal and operations. This role, however, was limited by the financial resources of these groups, but could be encouraged by such means as providing public funds for them to review existing analyses or to perform their own. A difficulty with this procedure is in deciding upon the groups that should receive the funds. An alternative might be to fund "Science Shops",[4] or groups at universities to provide technical aid and advice to citizens' groups. Either of these methods would be a step towards encouraging more critical review of risk analyses carried out by industry or government agencies.

A difficulty with the above procedures are the inevitable differences in opinions that will result from simultaneous studies. Ackerman *et al.* (1974) point out that the traditional approaches such as legal responses, agency hearings, and judicial reviews have inherent limitations with respect to evaluating these conflicting assessments. As a way of dealing with this problem, they advocate the need to establish rules of evidence for scientific studies used in legal proceedings. These rules would encourage more uniform analyses, so that the debate could focus on the alternatives themselves rather than the particular assessment or presentation promoted by an interested party. Lathrop and Linnerooth (1982) provide a suggested set of guidelines with respect to establishing rules of evidence, and stress the importance of defining the risk being assessed, clarifying the assumptions and error bounds, as well as indicating the conditional nature of specific analyses that are undertaken.

Decisions also have to be made in each country as to when specific analyses of potential negative effects should be undertaken. There were clear institutional differences between the four countries on this point. In the UK the Secretary of State, after granting planning permission in principle, stipulated that a full safety audit should be undertaken prior to the commissioning of the facility. The difficulties over timing in this case, however, have led to changes in the procedure for handling future planning applications in that new applicants now have to submit more detailed plant designs earlier in the process. In the US a number of different analyses were undertaken soon after three sites were proposed for California. In both the FRG and the Netherlands analyses were undertaken prior to the approval of the projects, but these studies did not play a central role in determining the final outcomes.

There is no easy answer to the question as to when analyses should be undertaken. At the LEG Task Force Meeting in September 1980, the following contrasting views on this point were expressed by Ralph Keeney, a decision analyst from Woodward Clyde Consultants in California, and Robert Norton, a manager from Distrigas in Boston, Massachusetts:

> *Keeney*: Responsibly, siting analysis should be done before the decisions on the site are made, and this is personally where I would really like to see analyses done. (KLS 1982, p369)

> *Norton*: I think it may be useful for existing terminals, but I do not think it will prove useful for siting purposes . . . . The real benefit in risk analysis is when it is relatively constrained, where you can work within the system, namely an existing plant. (KLS 1982, pp289–90)

Perspective 2: Eliciting preferences

Analysts have proposed several different techniques for eliciting preferences and values of individuals with standing before a decision is made.[5] Wynne (1982) contends that it may be difficult for individuals to articulate their preferences because their value structures are open-ended, and tentative. The approaches outlined below for eliciting values must be interpreted with this potential caveat in mind.

The best known methodology is *multi-attribute utility analysis*, a formal technique for eliciting different attributes and determining their relative importance with respect to a particular problem.[6] It is normally utilized in conjunction with just one interested party, although this stakeholder may have obtained inputs from other groups concerned with the final outcome. For example, three analysts (de Neufville, Keeney, and Raiffa) assisted the Mexican Ministry of Public Works in selecting a strategy to develop airport facilities for the Mexico City metropolitan area. With the assistance of these analysts, the Ministry estimated the utility functions for each of the relevant attributes, as well as the weights that would be assigned to them. In undertaking these assessments an attempt was made to incorporate the concerns of all interested parties (e.g., residents near the airport, the ministry of Communication and Transport) by utilizing information from previously commissioned reports on the airport problem.[7]

Recently there have been two other promising approaches that may enable analysts to obtain quantitative data on the preferences of different interested parties. One of these is *value tree analysis*, developed by Edwards and von Winterfeldt (1981), in which individuals or groups are presented with alternative scenarios regarding the choice between several different technological options. On the basis of these scenarios the analyst attempts to build a value tree to capture all expressed relevant corners. Prototypical value tree analysis consists of a set of risk—benefit dimensions that are then operationalized through a set of measurable variables. Through the use of traditional multi-attribute utility analysis techniques, importance weights are assigned to the different attributes comprising the value tree.

Another systematic procedure is the *analytic hierarchy process* proposed by Saaty (1980), which represents the elements of a problem through a hierarchical structure. Figure 9.1 illustrates the elements of this approach in the context of LEG facility siting. The first level of the hierarchy is the single overall objective: which LEG site (if any) to select. In our case studies, with the exception of the US, this question was decided by industry and did not reach the political arena. The second level specifies the set of five interested parties and then lists the attributes that are considered to be important to each of these groups (level 3). The lowest level consists of the set of alternatives available at a given point in time. Priorities are established within each level by assessing the relative importance of one element over the other in a pairwise comparison with respect to the criterion in the next higher level. For example, the importance of each interested party with respect to the others (level 2) will be determined in reference to the question of siting a facility (level 1).

It is difficult to see how these techniques would have been useful in the four LEG case studies within the existing institutional arrangements. All interested parties would have had to feel that trade-offs between attributes were possible and that it was desirable to make such comparisons. Even if

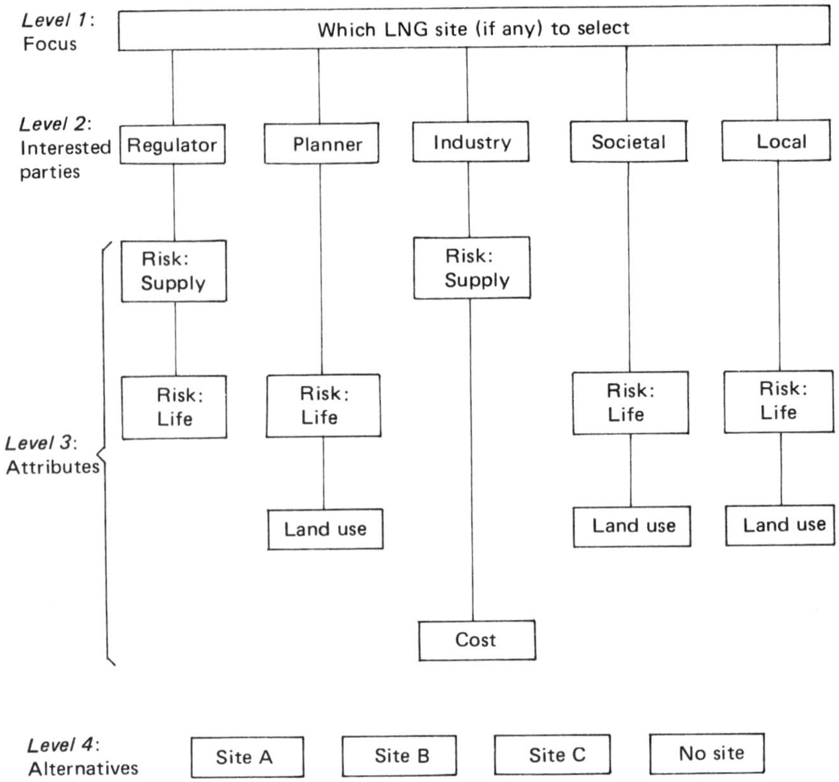

Figure 9.1. Hierarchical structure for siting an LEG facility.

the parties knew their value structure they might have been reluctant to articulate it for fear of losing some bargaining and negotiating power. As an analyst, Ward Edwards encountered this problem in evaluating alternative school desegregation plans submitted to the Los Angeles School Board. He noted that the interested parties in a societal decision problem may be unlikely to reveal their value structure because when this information becomes public certain groups might feel they would be held accountable for their judgments (Edwards 1981).

This concern was also voiced by William Ahern of the California Coastal Commission, at the LEG Task Force Meeting, when he responded to his own rhetorical question, "What was the role of analysis . . .?" as follows:

> This is a perfect project for multi-attribute decision analysis. Five perfect sites. You can compare them on environmental risks, safety problems, the cost of the sites, the technical problems. It was a classic. In fact, I haven't told this next bit of information even to Ralph Keeney.
>
> Ralph works up the street from us, and he came over to our office, as I understand, and talked to some people on my staff and he asked, "Wouldn't you be interested in a nice multi-attribute analysis of these alternative sites?" Well, my staff came to me and said, "Bill, this guy is from Woodward-Clyde and, gee, he

> sounds good. Shouldn't we hire him to help us do this?" And I said, "Absolutely not!" And it's not because he isn't an excellent analyst, but because he doesn't work for me and therefore he would be out of my control in many ways, because Ralph talks to people such as yourselves. He's got his own principles to uphold, and he would not give us a report that we could put under the shelf because Norbert Dall would know that we had contracted with Ralph Keeney to do the report, and he would want to see it — no matter what it came up with, and that's not something we wanted.
>
> Not only did we on the staff not want it, but our commissioners didn't want it. Actually I'm trained in this stuff myself, and I tried to talk my staff into doing it as they, at least, work for me. And if they put a number in I didn't like, I could jiggle it around. They were so appalled at the idea of doing this. (KLS 1982, p284)

Ahern went on to point out that the commissioners were not interested in such an analysis because they did not want to be explicit about their reasons for making their decisions.

By eliciting preferences at an early stage in a siting process the analyst may be able to point out the types of conflicts that are likely to be difficult to reconcile, and those where some negotiation and bargaining may be possible. For example, if there are disagreements on social values so that one group feels that "small is beautiful", while another feels that "large is necessary", then this may suggest that some new, more politically acceptable, alternatives are required (von Winterfeldt and Edwards 1981). On the other hand, if one group feels that it is bearing the risks of a project without receiving sufficient benefits then this information may facilitate trading arrangements between groups. Some of the policy options that might be considered are discussed under Perspective 3.

Finally, one has to be aware that the preferences of different individuals and groups may change over time. In the context of the multi-attribute, multi-party (MAMP) framework there were clear changes from one round to the next. For example, in the FRG the environmentalists and citizens' groups only raised objections to the LNG terminal in Wilhelmshaven after learning that Imperial Chemical Industries wanted to site a new petrochemical plant in the area. In the US the Sierra Club initially favored Oxnard as a potential site but later reversed its position.

This does not mean that analysis cannot be useful in shedding light on preferences, but it does suggest that what is viewed as desirable today may not be deemed appropriate several years later. Michael Thompson pointed this out with an interesting account of a housing project at a British university:

> When taking individual values or preferences into account, one of the troubles is that people don't stay in the same place. Sussex University went to a terrible lot of trouble to find out what students would like, in particular what kind of buildings they would like. It took several years for the buildings to go up and as the students only stayed there about three years, none of the students whose preferences had been taken into account were there any more. There were a whole lot of new students who detested the buildings. Furthermore, the same people change their minds, change their preferences, so even if you haven't got a fresh load of people

there, you often find that their values have changed — in some cases dramatically, which we see in the US, with the changing attitudes to nuclear power, for instance. (KLS 1982, p365)

Perspective 3: Designing proposals

Policy analysis may also be useful in developing proposals for selecting the location(s) for new facilities, as well as formulating options for distributing the potential gains and losses from final decisions.

Selecting a community. The four case studies suggest several different procedures for selecting a site for a hazardous facility. In each country the discussion of a possible LEG terminal was initiated by a utility company who felt that a new plant would be economically viable and would be in the interest of the nation's energy supply. In the FRG, Ruhrgas and Gelsenberg proposed Wilhelmshaven because the *Land* authority (Lower Saxony) had set out to encourage industrial development by deepening the shipping channel and by reclaiming from the sea a large area of land designated for industrial installations. As early as 1972 the two gas companies even established a subsidiary firm based in Wilhelmshaven to indicate how serious they were with respect to this site. In the UK a similar selection process was followed. Shell and Esso pursued the possibility of developing an LEG facility in the Mossmorran—Braefoot Bay area knowing, amongst other things, that there was considerable concern with the high level of unemployment in the area. These two examples illustrate one type of siting procedure: the developer negotiates directly with representatives of a community and attempts to work out a settlement.

In the Netherlands Gasunie attempted to follow this same process with respect to a Maasvlakte site, but ran into strong resistance on safety grounds from the Rotterdam local authorities. The company thus approached the Groningen local authorities about the Eemshaven site after they had marshaled strong political support for this location in order to aid industrial development.

In California, Western LNG Terminal Company proposed three sites for consideration and then attempted to gain approval for them. They were encouraged in their efforts by local authorities in Oxnard and Los Angeles, but met strong resistance from citizens' groups and environmentalists. This procedure illustrates a case where various interested parties react to a set of options initiated by a developer.

In judging the relative merits of these three procedures, one important consideration is the cost in terms of time and money in developing plans for a terminal and having to negotiate contractual arrangements with the relevant community. If there is great uncertainty on the part of the developer as to how long this process will take then there may be a reluctance to propose a site in the first place. This is the current problem facing the nuclear power industry: utilities are disinclined to initate new proposals for plants given the uncertainties associated with reaching licensing agreements. A second consideration is the importance of such projects in terms of national energy needs. If an LEG terminal is considered essential to the country then there is likely to be considerable pressure from government authorities to find a site.

An alternative procedure is illustrated by the Massachusetts Siting Hazardous Facility Waste Act of 1980 (O'Hare *et al.* 1983), under which the community has entitlement to the land and the applicant is required to negotiate a satisfactory settlement. Members of the community are represented by a local assessment committee;[8] surrounding communities can act through a siting council to have specific costs imposed on the applicant as compensation for potential losses. The siting council can also force the negotiating process into binding arbitration if the interested parties are not responsive to each other. To date, however, no community has agreed to negotiate voluntarily on a potential site.

In the light of the Massachusetts experience, the following sequential process for selecting a site for an LEG terminal might be considered. First, a set of acceptable sites is selected and conditions are established as to how gains and losses could be distributed among the interested parties. Second, a lottery could then determine which site would be selected. Any community not picked from the lottery could then make a more attractive offer to the developer in the hope of being chosen over the site selected at random. One type of process that may be relevant at this second stage has been proposed by O'Hare (1977), who suggests that any community interested in having a terminal in its backyard can determine a minimum level of compensation for which it would be willing to make a legal commitment to accept the project. The compensation could be in monetary form (e.g., taxes) or in kind (e.g., having the developer allocate some of this land for a park). If no community wanted to enter such a bidding process, then the site selected by the lottery would be declared the "winner".

Sharing gains and losses. The choice between alternatives can be facilitated by developing programs for sharing gains and losses from a proposed project. At a descriptive level, policy analysis can indicate what is likely to happen if a certain course of action is followed, while from a prescriptive point of view it can recommend what should be done to achieve specific objectives.

To illustrate these two aspects consider the history of nuclear power in the US. At a descriptive level, risk analyses by the industry revealed that the probability of a major nuclear accident was extremely remote, but that if it did occur the consequences would be very serious. Private companies felt that even a remote chance of a catastrophe would be a major setback to their participation in the development of nuclear technology. The exposure of firms to large liabilities also increased the awareness of the public to potential losses. If there were an interest in the development of nuclear energy then the government felt its involvement would be necessary to encourage firms to participate. Furthermore if there was a desire to compensate the public for damages caused by a catastrophic accident then federal relief would be needed (Chevarley 1975). These analyses led to the passage of the Price—Anderson Act of 1975, which established a complex program of private financial protection through a nuclear liability insurance pool and government guarantees covering total claims up to a maximum of $560 million in the event of a nuclear accident.

Turning to LEG, there has been no effort by any of the four countries under study to establish special insurance programs for encouraging industry to develop these facilities. The energy supply companies have not voiced undue concern about their potential liability should a large-scale accident occur. In fact, it appears as if the applicants and the insurance industry feel that the prospects of their having to deal with the consequences of an accident are sufficiently remote that they prefer not to address the issue. Only citizens have recently become alarmed when a facility has been proposed in their own backyard.

The proposals in this section involve types of compensation for individuals or groups. A distinction can be made between *ex ante* compensation, which refers to payments in money or in kind at the time a terminal is approved or constructed; and *ex post* compensation, which refers to reimbursement to individuals or groups who suffer losses from an accident.

Ex ante compensation: If considered appropriate on social and political grounds, *ex ante* compensation should only be given to those parties who can make a convincing case that they stand to lose as a result of a decision to site a new facility. Those individuals who have entitlement to the land will demand a certain amount of money before they will relinquish their property rights. However, if government exercises special powers to acquire this land then these individuals may not be satisfied with the settlement. Life becomes even more difficult for residents near the site who have no formal entitlements to the property on which the facility will be located. One way to compensate them could be through reduced tax rates.

One proposal with potential merit is an agreement by the applicant to reduce electricity rates to residents within a certain distance of the hazardous facility in order to compensate them for the increased risk or unpleasantness created. Such a system has recently been introduced in France with respect to nuclear power plants. People living within approximately 15 km of a facility who feel that they are adversely affected can apply to the local authority for a reduction of up to 15—20% in electricity rates. This compensation applies to both businesses and private households.[9]

The most difficult problem in designing *ex ante* compensation schemes is to distinguish between those who will bear direct losses, and those who are resisting the development for more general reasons concerning the future of society as a whole. This distinction is a subtle one, particularly when it involves questions of safety and environmental impacts that affect individual members of the community, as well as future generations. Linnerooth (1982) suggests that policy makers should consider the possibility that the public may wish to distribute equitably the costs and risks of industrial society. Yet, monetary payments to groups who resist a project, but who will not be directly affected by it, may encourage others to demand a share of the pie. In addition, society wants to preserve the belief that life is special; money may cheapen it, while laying bare the inequality of wealth. Calabresi and Bobbit (1978) make this point by noting that "the willingness of a poor man, confronting a tragic situation, to choose money rather than the tragically scarce resource [life] always represents an unquiet indictment of society's distribution of wealth." The authors contend that a democratic society should not tolerate such exchanges, even if both rich and poor prefer them.

For these reasons monetary compensation to protesting groups has rarely been implemented. One example we have found is from the FRG, where the utility company STEAG (Steinkohle-Elektrizitätswerke AG) announced plans in 1976 to construct a 1400 MW coal power plant in the city of Bergkamen in the Ruhr area. A citizens' action group protested the project and threatened to delay the licensing procedures. In March 1977 a contract was signed between the utility company and three representatives of the action group: the group would be reimbursed with a DM 1.5 million payment if they agreed to cease protesting the project further. However, a court case was provoked when the city of Bergkamen refused to distribute the money. The federal court decided that the contract was valid because the citizens' group should be compensated for legitimate rights, but the decision was greeted with negative reactions by FRG public opinion. Concerns were voiced in the media that health and safety were citizens' inalienable rights that could not be paid off with money (Blount 1983).

Compensation in kind or for specific causes appears to be a more acceptable way of sharing the pie than direct monetary transfers to individuals. In this spirit, proposals have been made that communities should be provided with specific facilities such as a hospital, which reduces the risks in other spheres of daily activity. In the FRG, the increased subsidy provided by the state of Lower Saxony for the recreational facility at Hooksiel is a good example of how this type of compensation system was implemented in the context of our case studies.

Another example of this type of compensation relates to the construction of a 1500 MW coal-fired power plant in Wyoming. A law suit had enjoined construction of the plant because of its potential damage to the surrounding environment. The suit was settled when the utility companies agreed to set up a $7.5 million trust fund for the express purpose of preserving a 60-mile stretch of the Platt River, the habitat of migratory birds, including the whooping crane. The coal plant was completed in 1981 and is fully operational today.[10]

Ex post compensation: These proposals require a set of *liability rules* that transfer entitlements from one party to another in the event of an accident. In the case of hazardous facilities such as LEG terminals, the principal question regarding this type of compensation is who is liable for losses should an accident occur. Pfenningsdorf (1979) points out that for ultra-hazardous or abnormally dangerous activities, public policy supports the doctrine of strict liability whereby the operator of the facility is liable for damages, regardless of fault. Accidents associated with LEG vessels and terminals fall under this category in all four countries we have studied.[11]

Both the applicant and the potential victims are naturally concerned as to how they will fare after an LEG-related accident. From the point of view of the applicant, a large-scale disaster may cause bankruptcy if the firm were required to pay for the entire loss, and so it may decide not to build a facility. Residents may not know how much they will receive in compensation for property damage they may suffer. Some may feel that no amount of money could compensate them or their surviving relatives for loss of life or severe injuries; they may thus oppose the facility.

If the society feels that *ex post* compensation for damages from accidents should be entirely the responsibility of the developers of new technological facilities, then private or public insurance appears to be a policy tool worth considering. Insurance would have the advantage of creating incentives for firms to make their facilities safer. If premiums are based on the risk associated with a given design then the applicant may want to spend extra funds to reduce the chance of an accident and correspondingly lower the annual premium for a given amount of coverage. The applicant also has an incentive to hire experts to estimate the risks associated with the facility, which can then be used as inputs into the negotiations.

The most difficult problem associated with developing a meaningful insurance program is the uncertainty associated with the risk of a catastrophic accident. Private insurance and reinsurance firms are concerned with the magnitude of the losses from a low-probability event, so they are reluctant to enter this market. We could not determine in any of the four case studies what proportion of the damage from a disaster would be covered by firms, how much the goverment would pay, or what proportion would have to be absorbed by the victims themselves. These distributional questions should be more explicitly considered as part of the siting decision process itself, rather than after an accident has occurred.

Another form of *ex post* compensation are payments in kind to groups that have been adversely affected by a particular disaster. A landmark case in this spirit was the settlement by Allied Chemicals in Virginia after being found guilty of polluting the James River with the pesticide Kepone. Rather than paying a \$23.2 million fine, the company proposed paying \$5.2 million and establishing an \$8 million trust fund to be used for environmental grants in Virginia.[12] In essence, the firm provided *ex post* compensation for research to prevent future damage to the environment.

The other extreme position is when society feels that *ex post* compensation for damages is primarily a public responsibility. In this case, the government could provide disaster relief in the form of low-interest loans or grants. Alternatively, some type of damage compensation fund could be established to aid victims. For example, the government could impose a tax on applicants to obtain funds for operating such a program to handle losses. The public sector would thus have a stake in the operation of the facility and it could thus set a tax either to encourage or discourage certain types of operations (Okrent 1982, personal communication).

Summary. These proposals for sharing gains and losses are designed to facilitate the bargaining and negotiation process between the interested parties who have standing in a siting debate. If certain individuals have the entitlements to block the approval of a facility that promises to increase general social welfare, then society may want to consider some of these proposals as a way of sharing the benefits more equitably. Considerably more empirical research is needed to determine how well each of these proposals is likely to perform in different situations.

How difficult will it be to implement specific plans as the population becomes more heterogeneous? What are the costs associated with implementing and enforcing certain property and liability rules? How acceptable are certain proposals likely to be within consensual and adversarial siting processes? These questions are beyond the scope of this book but suggest avenues for future study.

NORMATIVE CRITERIA FOR A DESIRABLE SITING PROCESS

This section recommends selected normative criteria that might be explicitly addressed in determining a siting procedure for any given country. These criteria are couched in relative rather than absolute terms since each society will determine its own appropriate targets. The status quo frequently serves as a benchmark for specifying these criteria and it may not be viewed as desirable to stray far from existing procedures. For example, Niall Campbell from the Scottish Development Department takes a conservative view that only incremental changes should be made in the public inquiry system in the UK, based on the experiences in the Mossmorran—Braefoot Bay siting decision:

> The way to make the inquiry system better is to change small parts of it. I know that is not a very exciting conclusion. You could examine the time limit, you could look at the amount of advance preparation, how much in advance, and how extensive it is. It is the degree that we are looking at rather than fundamental changes. (KLS 1982, p236)

By specifying these criteria one can then design a set of rules to satisfy them. For example, in designing rules of the road there are trade-offs between permitting drivers to reach their destinations as rapidly as possible and society's concern with highway safety.[13] The criteria described below relate to the decision process itself, as well as to the outcomes resulting from the party interactions. The criteria thus address questions of procedural as well as substantive rationality. The final choice will be determined by each society on the basis of current and desired institutional arrangements.

Criterion 1: Degree of openness

How open should the siting process actually be? What should be the degree of participation of the different viewpoints in the siting debate? At one extreme is the philosophy that it is useful to hear the widest possible range of viewpoints before making a final decision, while at the other end of the spectrum is the position that the process should be highly centralized within the government bureaucracy.

The first position is exemplified by the Berger Inquiry, which explored the technical, social, economic, and environmental issues of building an oil and gas pipeline through the Mackenzie Valley in northern Canada. Justice T.R. Berger, who headed the inquiry, traveled 17 000 miles in the Northwest and Yukon Territories over several years to hear evidence in towns and villages. The Inquiry offered funding to numerous native and regional organizations to enable different groups to provide testimony. Organizations were funded who had a clearly discernible interest that Berger felt should be heard but who could not afford to present their positions on their own. All groups had to have clearly delineated proposals as to how they intended to make use of the funds. Government and industry spent millions of dollars on the process (Gamble 1978). In the end they decided not to build the pipeline.

The other extreme, centralization, is illustrated by the French system. Louis Vincent of Gaz de France has pointed out that discussions regarding siting take place predominantly inside the ministries in Paris. With respect to the construction of LEG terminals, there is only one government agency in France with which Gaz de France has to deal, and no public interest groups have voiced concern about proposed projects (see KLS 1982).

In the four countries we have studied, there are clear differences in the degree of openness of the siting processes. The systems in the Netherlands and the US allow considerable room for debate and discussion between interested parties, but the other two countries have more structured arrangements: in the UK different points of view are aired through public inquiries, while in the FRG, through discussions between the utility companies and the licensing authorities.

Points can be made both for and against a more open process for participation in siting decisions. The presentation of a wide range of viewpoints enables all parties to feel they are part of the process. New information can change the outlook of some groups and may suggest a wider set of alternatives than had previously been considered. An open process may thus create possibilities for negotiating settlements between parties by suggesting ways of redistributing some of the gains from winners to losers (Orr 1977). On the other hand, such a process takes time, is costly, and may create additional conflicts between parties as certain issues in the debate are explicitly raised.

Ultimately, cost will influence the degree of openness. Only affluent countries can afford to have such time-consuming decision processes as the Berger Inquiry, which at the time was considered a landmark study with respect to the degree of public participation. It is doubtful, however, whether governments and industry in any country will invest this magnitude of funding within the near future to facilitate open siting processes.

Criterion 2: Nature of deadlines

How rigid should time schedules be with respect to different phases of the siting process? Should there be considerable flexibility, or should deadlines be specified in advance and revised only if an interested party can present a convincing argument for delaying the process? The Berger Inquiry in Canada is an example of a process that was somewhat open-ended. Justice Berger felt that hearings should be continued until all parties had had an opportunity to present information that they felt was relevant to the pipeline decision. As a result the Inquiry lasted three years, from March 1974 to March 1977 (Gamble 1978).

As a contrast, the Windscale Inquiry in the UK into the need for and desirability of a nuclear fuel reprocessing facility ran only 100 days. In fact, it represented for the first time in British political history that a technological debate had been rigorously structured. Mr. Justice Parker finally recommended that planning permission be granted to British Nuclear Fuels without delay, despite serious objections by public interest groups that it would be unwise to do so (Wynne 1978).

Deadlines formed an important part of the LEG siting process in all four case studies. In both the FRG and the Netherlands the contracts with the Algerian company Sonatrach stipulated that the exact locations of the sites for an LNG terminal be specified by October 1978, and this spurred the decision processes in both countries. In the UK the key deadline was the conclusion of the public inquiry, after which additional evidence could only be considered if the Secretary of State felt it was relevant. In the US the California LNG Terminal Siting Act of 1977 stipulated that a final decision on a site had to be agreed upon by July 1978, at which time Point Conception was conditionally approved.

One of the principal arguments for setting firm deadlines is that the applicant can plan its investment strategies with greater certainty. Fixed schedules also reduce the incentives for other interested parties to use information as a delaying tactic (O'Hare 1981). In the case of LEG there are important international implications of failing to meet agreed deadlines, as illustrated by the following comment by Philippe Cruchon of the French Ministère de l'Industrie:

> Generally, when a contract is signed with an importing country, this contract specifies a time schedule for the fulfillment of the contract. And there is a link in the contract between the construction of the receiving plant in the importing country and the construction of the liquefying plant in the exporting country. From an importing country's point of view, the contract is not only a private business matter, but it is also a matter of the international credibility of the consumer country. (KLS 1982, p435)

There are also advantages of not setting specific deadlines, in that parties do not become locked into a particular technology or specific site, and hence are less constrained in dealing with new information. Norbert Dall, project manager of the Sierra Club from 1976 to 1980, reflected this point of view when commenting on the siting process in California:

> The very assumptions of the project were critically reviewed by government agencies and interested parties that, in part, did not share the applicant's point of view. That process will, as far as the Sierra Club is concerned, continue as long as the gas supply–demand balance changes in California. As such, the review process is a clear expression of American pluralistic politics . . . . Thus we have perceived the present expenditures to review the project as a prudent investment that may keep California from making an economically and environmentally horrendous mistake. (KLS 1982, unpublished draft)

It can be concluded that parties or organizations that support, and may depend on, economic development will want well specified deadlines. Those who prefer to reflect on their technological future will insist on a more open-ended and flexible process.

Criterion 3: Specificity of contractual arrangements

How detailed and well specified should contractual siting arrangements be? Is it desirable to indicate in writing exactly who is responsible for damages from specific accidents, or should these arrangements be left intentionally vague? These questions came to the fore after the Three Mile Island accident, when there was a problem as to who would assume actual financial responsibilities for damage: the utilities, insurance companies, or the federal government. Similar questions are likely to arise should there be a catastrophic LEG-related accident.

In addition to questions of responsibility for an accident there are other consequences of the siting decision that may affect local residents. What happens if property values fall as a result of locating a hazardous facility in the area? Will these individuals be compensated for their loss in net worth? Suppose there are unexpected negative environmental impacts from a facility. Do residents of the area have legal recourse? Opposition to a proposed facility from citizens may arise because individuals do not have information on what recourse they have to recoup these losses.

Contractual arrangements should be explicitly specified so that all parties have a clear idea of their entitlements and those of others when agreeing to a particular siting proposal. This would also provide a firm basis for negotiations between groups in the siting debate since the potential gains and losses would be more clearly delineated. In some cases, less explicit agreements may facilitate the siting process by not raising a set of concerns by some of the parties. For example, if an actual dollar figure were attached to a severe injury or loss of life from an accident then people may disagree on the amount, and considerable time might be spent debating this point. On balance, however, it is important for all parties to know who has property rights to the site in question and who is responsible for damages should an accident occur after a facility has been built.

Criterion 4: Nature of compensation

What types of compensation systems, if any, should be introduced into the political arena for distributing the costs and benefits among the affected parties? Should there be an explicit recognition as to who are likely winners and losers, or is this an issue that society prefers not to address?

The philosophical underpinnings guiding societal decisions will influence the role that compensation plays in the siting process. For example, a libertarian system, which forms the basis for the operation of a free market system, stresses the importance of individual freedom unless others may be harmed by certain actions. The Pareto criterion is used as a guide to judge future actions. If the status quo is the baseline from which to judge future actions, then any project that is likely to harm a single individual would not be approved unless that person is sufficiently compensated that she will consider herself at least as well off after the decision is made as before. A utilitarian system, on the other hand, is based on the objective of maximizing the utility of society as a whole. Using this ethical system projects would be approved even if some parties would be made worse off than under the status quo.[14]

The issue of compensating potential losers is closely related to the question of the rights of individuals and groups with respect to their safety and quality of living. If individuals have a right to reject a project because it imposes environmental and safety risks on them, then the compensation issue is likely to play a central role in dealing with these problems of risk. If certain risks are imposed on individuals without their consent then issues of compensation assume a less important role.

The four case studies illustrate different philosophical positions regarding the issues of rights and the role of compensation. In each of the three European systems a site was approved because there was sufficient positive feeling by local and regional groups that the project would provide added employment and revenues to the area — compensation in kind rather than in money. In the US, the city councils in Los Angeles and Oxnard also favored the sites for the same reasons, yet the Oxnard Citizens' Group had sufficient standing that their concern with safety risk issues was instrumental in restricting potential sites to more sparsely populated areas than Oxnard. The Joint Action Group from Aberdour and Dalgety Bay in Scotland was also interested in the safety risk issue, but felt that authorities in the UK did not listen to their concerns. The rights of citizens in the FRG are protected through the courts, and the residents of Hooksiel were able to obtain additional subsidies for a planned recreational center because of their concerns that the LNG facility, together with the adjacent petrochemical plant, would have a negative impact on tourism in their area.

Providing opportunities for trade and negotiation between parties may help to reduce conflicts by pacifying individuals and groups who would otherwise be opposed to a project. Negotiating compensation agreements can be a time-consuming process and it may be difficult to design a system in which some groups do not behave strategically in an attempt to obtain more than they actually require or deserve.[15] Furthermore, there may be differences in preferences between individuals in the same political jurisdiction (e.g., a local municipality). In such cases some individuals will be more than satisfied with a proposed compensation scheme, while others may feel they have not been offered enough to make the project attractive. There may also be a concern that letting tastes enter into the choice of sites may be inappropriate.

Recognizing the procedural difficulties, compensation to those who stand to lose from a new facility siting decision may be desirable if this action also improves social well-being. In this case, compensation would be consistent with the Pareto criterion: if you can make everyone better off by transferring some gains from winners to losers, then this is a desirable course of action to follow. Compensation will be most appropriate in countries where the decision process is based on a compromise model. Under a judicial model, where a number of policy makers have final responsibility for a decision, then it is not necessary to reward certain groups in order to reach a final decision. In this case the parties responsible must weigh the relative importance of welfare and distributional objectives in making their final choice.

Louis Clarenburg pointed out the dilemma faced by the Dutch on this question:

> A major problem that we now are facing is how to weigh the voice of a small population against the total population when you are talking about projects of national interest, the problem of equity. At what level should decisions in such cases be taken — at the municipal level, which has to take into account all the interests of their own population, or at the provincial level, or at the national level? That is the question which remains in my mind and which I haven't solved. (KLS 1982, p264)

We have not solved this problem either. A siting decision represents a balance between political and economic considerations. To improve the policy process, policy makers and analysts need to have fruitful dialogues.

CONCLUSIONS

Finding a site for an LEG terminal, or deciding not to site the terminal, is as much a problem of determining how to distribute the gains and losses of technological society as a problem of how to increase economic welfare. The way in which these trade-offs were made depended on the institutional arrangements of the four countries, which followed to varying degrees the judicial or compromise models for making siting decisions. An important conclusion of this chapter is that analysts *can* help by assisting decision makers on problems involving *both* welfare (or efficiency) questions as well as distributional questions.

The analyst, in his or her more traditional role, can clarify the potential effects of a proposed facility. By eliciting preferences of the contending parties, the analyst may be able to point out those conflicts where some negotiation and bargaining would be possible. Finally, policy analysis may be useful in developing proposals for selecting the sites, as well as formulating options for sharing the potential gains and losses of those affected.

The problem of siting an LEG terminal is not, however, only one of resolving the welfare and distributional objectives, but also encompasses an equally complex set of questions concerning the *process* by which the decisions are made. Four possible criteria have been suggested here that should be addressed in determining a siting procedure: the degree of openness, the nature of deadlines, the specificity of contractural arrangements, and the nature of compensation. An open process is in many ways desirable, however costly. Deadlines expedite the decision process, but, again, have a cost in terms of limiting the time for reflecting on the future of the technological society. In contrast to the costs of providing for an open process with flexible deadlines, there appear to be fewer economic or social costs of assuring that the contractual arrangements are clearly specified so that each party has a clear idea of their entitlements and those of others when agreeing to a particular siting proposal. Likewise, compensating those who must inevitably bear the risks of technological developments appears to be an attractive policy recommendation, but must be considered within the political and institutional context of the country involved.

NOTES

1. See Chapter 2, p18–19 for a more detailed discussion of these concepts.
2. This point is discussed at length by Michael Thompson in the Postscript. In a similar vein, James Douglas (1983) has contrasted different political systems that have emerged to deal with conflicting preferences within the electorate. He shows how different institutional arrangements evade the constraints implicit in Arrow's (1963) four conditions of rationality, which imply the impossibility of eliciting social preferences from individual preferences.
3. Mark Pauly suggested the distinction between these two types of process models.
4. Science Shops are centers set up in Holland to distribute expertise, free of charge, to those who do not normally have access to it. They are staffed by university faculty members, on a volunteer basis, who provide information and undertake research and analyses for unions, environmentalist and neighborhood groups, etc., who are involved in negotiations (see Kunreuther and Ley 1982).
5. McFadden (1975) has developed a revealed preference approach for inferring the weights given to different attributes as well as individuals and groups after actual decisions were made by government bureaucracy. He and Phoebe Cottingham have applied this approach to an analysis of the California Division of Highways' selection of freeway projects.
6. For an excellent and comprehensive discussion of this methodology in eliciting preferences and value trade-offs, see Keeney and Raiffa (1976).
7. There was no reconciliation between two groups — the Ministry of Public Works and the Ministry of Communications and Transport. This conflict adversely affected the implementation process; although land was purchased for the new airport, construction of a new site has still not started.
8. This committee consists of the chief executive officer, chairman of the local conservation commission, local planning board, fire department, four residents appointed by majority vote of town officials, and not more than four members nominated by the chief executive's office and approved by a majority vote of the city council.

9. Gaz de France (personal communication, September 1982).

10. Personal conversation with Patrick Pateneau, Vice-President, Resource Conservation, US National Wildlife Federation, September 1982.

11. In the US a case arising from the death of a workman at an LNG site resulted in a court ruling that storage of natural gas in a populated area is abnormally dangerous because of its inherent risks (McLane *v.* Northwest Natural Gas Co., 467 P.2d 655 Oregon 1970). In 1976 New York enacted the Liquefied Natural and Petroleum Gas Act, which specifies that the storage, transportation, and conversion of LNG and LPG within the state are considered to be hazardous, thus entailing strict liability on the part of any person who undertakes such activities (US General Accounting Office 1978).

12. Virginia Environmental Endowment, 1977 Annual Report, Richmond, Virginia, USA.

13. Wittman (1982) provides a very interesting analysis of how rules for highway safety and sports reflect specific criteria.

14. See Schulze and Kneese (1981) for a more detailed discussion of the implications of different ethical systems on project selection and questions of public safety.

15. For a more detailed discussion of the strategic aspects associated with compensation, see Raiffa (1982).

Postscript

A Cultural Basis for Comparison

Michael Thompson

One thing that California and the UK have in common is that they have both approved sites for LEG terminals. After a long, drawn-out process in which it proved impossible to approve any of the proposed sites California finally, with the help of a new statute passed expressly for the purpose, was able to give approval for the remotest of all the sites on the list of possibles — Point Conception.[1] California's LNG Terminal Siting Act requires that, within one mile of the perimeter of the site, the population should not exceed ten persons to the square mile and that, within four miles of the site, the density should not exceed 60 persons to the square mile. Moreover, these stipulations also apply to the tankers laden with liquefied gas, which may be conceived of as mobile sites carrying their zones with them as they make their approach to the terminal or shelter offshore waiting for calmer weather before docking.

Scotland has a longer coastline than California and most of the country is very sparsely populated (less than 25 persons to the square mile) and yet the approved site, at Mossmorran—Braefoot Bay on the Firth of Forth, lies within the most densely populated part of the entire country (with a population density of 250—500 persons to the square mile). On top of this, laden tankers will pass within a mile or so of Burntisland (an industrial town), and sometimes within four miles of Edinburgh — the capital city of Scotland. If the California siting criteria (explicit in the 1977 Act) had been applied to the Scottish case it would have been quite impossible to approve the Mossmorran—Braefoot Bay site and, if the UK criteria (implicit in the Mossmorran—Braefoot Bay approval) had been applied to the California case, any of the suggested sites could have been approved, which means that the terminal would have gone to the first site to be suggested — Los Angeles harbor.

Surely, a glaring contrast like this can mean only one thing — that one (at least) of these two countries is way off-beam in its handling of the risks associated with LEG. But this "single-answer" approach has to make a rather dubious assumption — that Californians and Scots are the same (or, at least, that such differences that may exist between them have no bearing on the safe handling of these technological risks). If we reject this assumption then we can entertain the possibility that risk handling (and, indeed, technology generally) may vary with what in the eighteenth century would have been called "the temper of the people".

If this is the case then we cannot simply jump to the "single-answer" conclusion. We will have to entertain the possibility that, widely divergent though they clearly are, the Point Conception and Mossmorran–Braefoot Bay siting decisions may nevertheless each be *appropriate* to their different *contexts*. This approach would require us to focus not on the risks alone, but on the relationship between the risks and their social setting.

THE PROBLEM

When we look at our environment we do not see it with the naked eye. We see it as it is filtered through a cultural screen — our *idea of nature*. Yet the environmental theory of perception rightly points out that it is not what is inside our heads that matters, but what our heads are inside of. What happens when we put these two together and merge the contextual approach with the notion of a cultural screen? Instead of a direct relationship between an organism and its environment we have an indirect relationship in which the interposition of culture makes it impossible for the environment to be directly comprehended. Inevitably we perceive nature always through a cultural glass and always darkly. For animals, who have little if any culture, this is of scant consequence, but for humans, who have a lot of culture, it is nearly everything. Culture opens up an enormous interpretive space between the human organism and its natural environment, and it is this space that transforms our ecosystem into what Kenneth Boulding calls our echo-system.[2]

Our cultural screen furnishes us with a way of seeing and, more importantly, with a way of not seeing. We act in the world on the basis of what we see and our cultural screen then filters out nearly all the feedback that would help us to correct our view and, at the same time, lets through nearly all that confirms it. In this way perceptual bias is inevitable and learning inevitably confirms that bias.[3] This is not to say that there are no limits to perceptual bias — that there are no constraints on the dimensions and shape of the echo-chamber — but only that there will be some bias and that it cannot be eliminated. To accept this is to raise formidable obstacles to comparison. If two cases are similar, is this because they are similar in nature or because of a fortuitous convergence through cultural screens? And, conversely, if two cases are dissimilar, is this because they are dissimilar in nature or because of a fortuitous divergence through cultural screens? Do the case studies reflect similarities and differences between the cases studied or do they reflect similarities and differences between those who studied the cases?

A POSSIBLE SOLUTION

To accept that the perception of risks is biased is not to claim that risks are simply imagined. Risks, one may concede, are inherent in the universe; it is just that, inevitably, they are perceived through a screen and that, depending, as it were, on the cultural tint of that screen, some risks will gain perceptual salience and others will be filtered out. The cultural–environmental theory of perception simply says that *risks are selected* and that, therefore, bias in perception is inevitable.[4]

But biases in risk perception, it appears, are not infinite, nor do they run away in every conceivable direction; they seem to be rather strongly patterned and quite small in number. If this is indeed the case then it should be possible to abandon the universalistic position, which has to insist that the echo-chamber does not exist, without going over to the totally relativistic position that ends up according equal plausibility and legitimacy to every imaginable bias.[5] There is a theory, cultural bias theory,[6] which argues in favor of this notion of *constrained relativism* and proposes that only those perceptual biases that are socially viable stand any chance of persisting through time but, rather than elaborate the whole of this theory here, I will try to follow the more empirical trail that has already been blazed by a political scientist: David W. Orr.[7]

In trying to make some sense out of the energy debate in the US, Orr has identified three distinct perspectives,[8] each of which is appropriate to a particular set of primary actors and with each of which goes a preferred style of governance and a distinct set of salient risks. Each perspective, moreover, gains its particular orientation from the distinctive way in which the problem is defined (it is here, in the credible ways of defining the problem, that the different ideas of nature come into play, but more of that in a moment).

In what Orr calls the *supply perspective* the problem is inadequate energy supply, the primary actors are the energy corporations, the preferred style of governance is *laissez-faire* (a minimum of government intervention), and the salient risks are those associated with economic disruption. In the *conservation perspective* the problem is energy waste, the primary actor is government, the preferred style of governance is Leviathan (a major role for government), and the salient risks are those associated with balance of payments, overseas dependence, and energy wars. In the *energetics perspective* the problem is social and cultural, the primary actors are the public (I would prefer to say the public interest groups), the preferred style of governance is Jeffersonian (one in which a participating citizenry blows the whistle on government), and the salient risks are technological accidents, resource exhaustion, and climatic change.

Even in this skeletal form, Orr's tripartite scheme provides a powerful corrective to some of the bias that is built into our LEG study. For instance, one of the "difficulties" encountered in the US case study was that, when the senior executives of the energy corporations were asked about risk, they kept on going off at a tangent and talking about things like supply interruption and contract insecurity — the economic losses that might result from bad weather or breakdown preventing tankers from loading or unloading on schedule, and the even worse losses that would be incurred if gas-producing countries like Indonesia or Algeria were to renege on their contracts. When the executives were finally steered back to the topic of risks to life and limb they soon lost interest. "Oh *those* risks", they would say, "I think it would be best if you spoke to our Safety Officer/PR man about all that!"

What is so interesting about this little anecdote is that it shows the way in which the terms of reference for the LEG study have been drawn up in such a way that the perfectly legitimate (and, according to Orr's framework, the perfectly appropriate) risk concerns of the energy corporations are rendered anomalous. These terms of reference are, in fact, derived from the risk concerns appropriate to the energetics perspective (or, perhaps, from a governmental worry about how it should try to deal with the ever-increasing demands that are being made on it from the energetics perspective). If we were to leave this bias unidentified then this whole study, whilst claiming to be interpretive and descriptive, would inevitably become strongly normative and narrowly prescriptive. The real prescriptions would go through covertly under the label "description" with the overtly prescriptive concluding chapter providing a few finishing flourishes.[9] How can we avoid doing this?

Single-problem and multiple-problem analyses

If the analyst himself "decides"[10] what are the relevant risks (or if his clients do this for him in their terms of reference) then he is committed to the single-problem approach. Clearly, this is what was tending to happen in the US case study and it is what actually happens in the risk questionnaires administered by psychometricians — the risks are given by the format of the questionnaire and the interviewee is allowed to say how serious or trivial they are to him. In the multiple-problem approach the analyst would explore (by informal guided interview or by participant observation) in order to find out what sorts of activities and agents the subject regarded as risky (and, indeed, whether the analyst's concept of risk had any comparable counterpart in the subject's conceptual scheme). Only then, when he had elicited the "home-made" risk model, would he feel justified in asking questions about the comparability and relative seriousness of the various risks that loom large in the subject's life. Nor could he bring himself to use a methodology that required the subject to attach a fixed total of weights to the lexicon of risks that he (the analyst) had identified, without having first ensured that none of the values threatened was conceived of as sacrosanct (and therefore infinitely weighted).

From all this it might appear that the multiple-problem analyst is so fastidious that there is just no chance of his ever getting to grips with the task and that, regrettably perhaps, there is no option but to persevere with risk analysis in its present single-problem mode. But is this really so? An approach is single-problem if the analyst defines risk for all his subjects and it becomes multiple-problem if he allows each subject to define the risk. This latter is what Orr has done. Far from defining what the risks in energy are, Orr has assumed that risks are selected and he has allowed each of his primary actors to define the risks that gain salience in that perspective. If the various positions within the energy debate depend for their very existence upon risk being plural in this way then, by Ashby's law of requisite variety, any approach that assumes that risk is uniform will be in trouble. Are we in trouble?

The case studies

All our case studies started off by identifying the "interested parties" and by sketching out the various "party perspectives"; so they certainly set out with the best of multiple-problem intentions. But there can be no doubt that, whatever happened in the early stages of their genesis, by the time they have all been put through the MAMP framework they have all been stripped of much of their variety. One sympathetic reviewer of our study has spoken of the MAMP framework as "forcing everything into one mold" and another, expanding on the same theme, has put it like this:

> . . . you seem to be imprisoned by the notion that all the various partisan participants in problem solving agree, during any one round, on a definition of the problem. Indeed, you mark the progression from one round to another by shifts (always unanimous?) from one problem definition to another. But it is a common characteristic of interactive problem solving that many, perhaps most, of the participants each carry a distinct version of what "the problem" is in their minds . . . . They are not working on any one given problem, nor do they think they are.

So, in this multiple-problem hypothesis, we see the MAMP framework — our reader's aid — as a repressive device that, trimming a bit off this distinctive problem definition here, and pulling a plank out of that partisan platform there, forces everything into the uniform single-problem mold.[11] Orr's pluralist framework, by contrast, would allow each perspective the freedom to define its own problem.

Yet, there is, surely, a sense in which it is valid to speak, as we do in

Chapter 8, of each "round" as being framed by a particular "problem formulation"; that formulation being the agenda, as it were, that is set by the "convening body" for the siting process. The "convening body", of course, seldom has a completely free hand in setting the agenda for the process, but, nevertheless, the fact that the other interested parties participate in that process can be construed as indicating that they have, to some extent, accepted that particular problem definition. But only, so the multiple-problem protagonists would argue, *to some extent*. Though they all sit down to the one visible agenda, they all bring with them their own hidden agendas.[12]

So the conflict between the single-problem and the multiple-problem hypotheses has to do with inclusiveness. Both can handle the visible agenda (the its-the-same-the-whole-world-over part of the story), but only the multiple-problem hypothesis can go on to handle the hidden agendas as well (the things-get-done-differently-in-different-countries part of the story). The multiple-problem hypothesis concedes that some ordering principle will be needed if the case studies are to be compared, and it concedes that the MAMP framework is a pretty effective ordering device, but it is just that order has had to be purchased at a price, and that price is the acceptance of a repressive single-problem approach. But this does not mean that all is lost. If the single-problem approach is inadequate then we should expect serious anomalies and paradoxes to show up in the case studies. If they do show up then we can restore the multiple-problem approach, and put it to the test by listing these anomalies and paradoxes and then calling on the cultural hypothesis to resolve them.

The cultural hypothesis

Figure 1 sets out the essential features of Orr's framework. Each of these perspectives is a whole package with each component playing its part. I have put the *ultimate energy sources* in a separate box in order to emphasize that each package is put together in such a way as to lead inevitably to the desired future. One consequence of this is that risk is never just risk but always "risk-for" (in the same way that history is always "history-for"[13]). The "risks-for" are the sticks (the sanctions) that are being used to drive the society towards the desired energy future, and towards the desired pattern of social relations that is perceived as accompanying that future. Risks, in other words, are selected in order to provide rationalizations for preferred patterns of social relations. That, given the inevitability of "risk-for", is the cultural definition of risk.

If risks are the sticks, what are the carrots — the positive inducements for society to move in the desired direction? Answer: resources. From the supply perspective we live in a world of *resource abundance*; from the conservation perspective we live in a world of *resource scarcity*; from the energetics perspective we live in a world of *resource depletion.*[14] But how

Perspective / *Variables*	*Supply*	*Conservation*	*Energetics*
The problem	Inadequate supply	Energy waste	Cultural and social
Primay actors	Energy corporations	Government agencies	The public (public interest groups)
Energy goals	Inexhaustible cheap energy	Near term: efficiency Long term: inexhaustible (but not cheap) energy	Decentralized solar-based society
Preferred style of governance	*Laissez-faire*	Leviathan	Jeffersonian
Value system changes requied	No change	Small (or gradual) change	Large (and sudden) change
Salient risks	Economic disruption	Balance of payments Overseas dependence Energy wars	Technological accidents Resource exhaustion Climatic change
Ultimate energy sources	Breeder/fusion	Conservation leading to breeder/fusion	Decentralized solar, wind, and biomass

Figure 1. Orr's framework.

do these contradictory convictions about the nature of resources arise, and how is it that, in one world, these contradictory convictions can go on and on existing? To answer these questions we have to resort to cultural bias theory and we have to underpin Orr's scheme, which is essentially an explanation in terms of goal-seeking (the goals being set by the evident self-interest of the primary actors), at the much deeper level of goal-setting (where one asks how it is that the actors can come to know where the self-interest in which they act lies).

First, let me generalize Orr's primary actors into three distinct social types: *entrepreneurs*, *hierarchists*, and *sectarians*. Each of these obtains his distinct social identity from the social context that he finds himself in and that he strives to maintain. Each context is defined by a distinct kind of organization: the *ego-focused network* for the entrepreneur, the *hierarchically nested group* for the hierarchist, and the *bounded egalitarian group* for the sectarian. I argue that this typology of organizations is exhaustive — that these are the only kinds of organizations that are socially viable.[15]

But there are two provisos. First the construction of ego-focused networks is a competitive business and, if there exist opportunities for economies of scale, some networks (those of the forceful, skillful, and lucky individuals) will expand at the expense of others (those of the timid, de-skilled, and unlucky individuals). The result is a bifurcation into two social types: the *entrepreneur*, freely transacting from the center of his extensive network, and the *ineffectual*, very much restricted by the pre-empting of his transactional options that results from the proliferating networks of the entrepreneur. Secondly, the deliberate avoidance of all three of these organizational forms can also be a socially viable strategy in certain circumstances (one of which is the absence of opportunities for economies of scale). Individuals who successfully follow this strategy constitute a fifth social type: the *hermit.*

So, although there are only three organizational types, their dynamic nature results in a total of five social types. These five can be conveniently mapped onto two axes of social context: *group*, which has to do with the extent to which an individual is incorporated into or is free from bounded social groups, and *grid*, which has to do with the extent to which an individual finds himself subject to or free from socially imposed prescriptions (in which case, like our entrepreneur, this may be because he is inadvertently busy imposing prescriptions on individuals in other social contexts). With each of these social types there goes a distinctive socially induced strategy that is all the time justified and sustained by a distinctive cultural bias. The social types, in other words, are not just *there*; they are continually in contention, and the whole thrust of this cultural theory is that policy debates (such as those that surround LEG) are best understood as particularly visible and focused instances of this continual contention (see Figure 2).

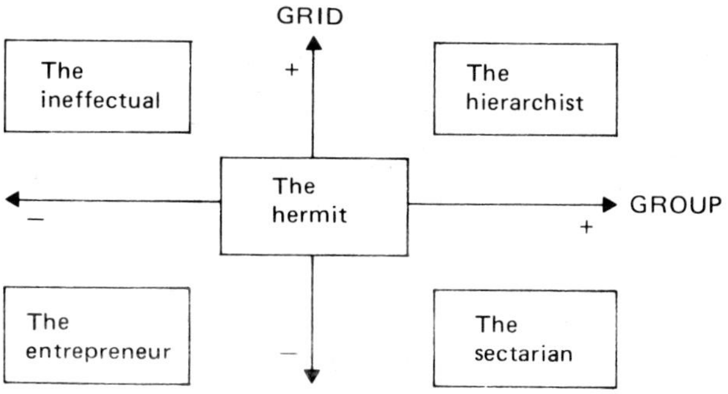

Figure 2. Social types and their social contexts.

The sectarian, the hierarchical, and the entrepreneurial biases are all clearly identifiable within our four case studies (though with some marked variations in their relative strengths, influences, and standings as we go from one country to another), but two social types — the ineffectual and the hermit — never show up at all. The reason is simply that the ineffectual could not gain access to the debate, even if he wanted to, whilst the hermit's whole strategy is based on steering well clear of all that sort of coercive involvement. Hermits drop out; ineffectuals are squeezed out. But, just because they are absent from the debate, it does not follow that they are irrelevant to it. Many public interest groups, for instance, credibly claim to speak on behalf of the poor, the helpless, and the disregarded. In so doing, they do not bring the ineffectuals into the debate; rather, they appropriate them as a moral bludgeon with which to belabor the Establishment — the entrepreneurs and the hierarchists — who together form the other diagonal of the social context map.[16] But more of these complex dynamics — these perpetual contentions — presently. First let me present the essential features of the three social types that do participate in the debate; then let me run through a number of anomalies and paradoxes within the case studies to see if I can resolve them with the help of this plural frame (see Figure 3).

Social types / *Variables*	*Entrepreneur*	*Hierarchist*	*Sectarian*
Organization (conceptual scheme)	Ego-focused network	Hierarchically nested group	Bounded egalitarian group
Cultural bias	Pragmatic materialism	Ritualism and sacrifice	Millenarianism/ fundamentalism
Socially induced personal strategy	Individualist manipulative	Collectivist manipulative	Group survival/ coercive utopian
Idea of nature	Skill-controlled cornucopia	Isomorphic (cornucopian with accountable limits)	Strictly accountable
"Carrot" justified by idea of nature	Resource abundance (culturally bestowed)	Resource scarcity (culturally bestowed within natural frame)	Resource depletion (naturally imposed)
"Stick" justified by idea of nature	Salient risks as in Orr (Figure 1)	Salient risks as in Orr (Figure 1)	Salient risks as in Orr (Figure 1)
Scenario that sticks and carrots are steering towards	"Business as usual"	"Middle of the road"	"Radical change now"

Figure 3. The cultural underpinnings for Orr's framework.

ANOMALIES AND PARADOXES

Rounds and bouts

Planning procedures (in market economies) create an asymmetrical arena within which the planning authorities act only when a proposer does something. Planning is defensive; proposing is offensive, and in the UK the initiative lies entirely with the proposer. The Shell/Esso consortium's aim is to obtain planning approval for a site of *their* choosing. No doubt they pay careful attention to economic factors such as pipeline costs and to environmental factors such as coastal configurations, prevailing weather conditions, and the Strategic Plan on which are marked the areas where development is variously encouraged or discouraged. If they are wise they will also pay some attention to social factors such as the likely strength of the opposition at the various points where they might try to put their facility ashore. And, if at first they do not succeed, then they can try, try, try, and try again.

In other words, the fight starts when the initiator decides he wants to put a facility ashore and it ends when he gets permission to do so (or when he gives up trying). From this perspective, the Mossmorran—Braefoot Bay proposal was just one round within a bout that began with Shell/ Esso's earlier abortive attempt to put their facility ashore at Peterhead. But the defense does not see it like that. It acts only when a proposal is initiated and it defines each bout as beginning with a proposal and ending with the approval or rejection of that proposal. Since there is no way in which these two perceptions can be reconciled within a single formulation, we really need two MAMP frameworks — one for the proposer and one for the planning authority — but we only have one! Let us briefly examine this anomaly by considering what would happen if we went multiple-problem and allowed plural MAMPs.

In the British case the mismatch between the two MAMP frameworks would be considerable; in the US case it would be less so. The reason is that, in the US, the need is debated first and then, if the decision is "yes", a list of possible sites is drawn up and the proposer has to work his way through this "public" list.[17] In Britain, the question of need does not get this sort of public hearing, nor is there a public list of sites (though, doubtless, the proposers have their own private list that they are prepared to work their way through). The comparison of mismatches between these sets of plural MAMPs is most instructive.

(1) Separate MAMPs for attackers and defenders permit us to recognize the fundamental asymmetry within the planning arena: the attacker's MAMP will reflect the entrepreneur's cultural bias, whilst the defender's MAMP will reflect the hierarchist's cultural bias. Accordingly, the pragmatic materialism of the attacker's bias leads him to treat the defender's non-negotiable preconditions (as set out in the Strategic Plan) as simply a part of nature, whilst the defender's bias in favor of clarity and order leads him to chop discrete sections out of the decision-making continuum at those points where he starts and finishes acting in response to the attacker's appearances and disappearances.

(2) The patterning of the mismatches between these MAMPs as we go from the US to Britain clearly reveals that things are much more "chopped-up" in Britain and that public participation is much greater in the US. This suggests a strong bias towards hierarchy in Britain and a strong bias away from hierarchy, and towards the sort of Jeffersonian demands associated with the sectist bias, in the US.

(3) These sorts of contrasts between Britain and the US are often generalized in terms of a distinction between *consensus* and *adversary* cultures, and the comparisons between plural MAMPs allow us to suggest a plausible cultural basis for the very different political regimes that underlie this distinction.

A representative but non-participatory style of democracy requires a deferential populace that is prepared to tolerate quite a high level of secrecy, that respects expertise, and that places trust in a system that divides issues up into morsels suitable for ingestion at various levels within its hierarchical structure. A more participatory style of democracy requires a truculent populace that is prepared to blow the whistle at the first sign of secretiveness, that suspects expertise, and that distrusts hierarchy-building tendencies whenever it sees them.[18]

So a comparison of plural MAMPs opens up a possible way of recognizing different political regimes and, through them, a way of getting at the different institutionalized styles of risk handling that go with them. In a *deferential regime* the balance that government has to strike is that between the demands that are being made from the entrepreneurial and the hierarchical cultural biases. In a *truculent regime* things are more complicated and the balance that government has to strike is a three-cornered affair that has to take account of the demands from the sectarian bias as well.

A reasonable prediction from this hypothesis is that a regime that demands public participation and opposes secrecy will generate much more literature (reports, transcripts of hearings, etc.) than one that restricts public participation and promotes secrecy. This is one comparison between case studies that can be made. By the simple criterion of weighing the "universes" of printed material accumulated during the course of the case studies, it would seem that all three European countries share consensus cultures whilst only the US has an adversary culture.[19] In exploring the remaining anomalies I will look, first, to see whether this tentative separation is supported and, second, to understand some of the unique properties of the European cases that impart rather different flavors to the one style of culture that they share.

Siting criteria compared

Only in the California case, with its LNG Terminal Siting Act, are the siting criteria made explicit. In the other three cases the criteria have to be deduced from specific instances of approval and rejection. Of course, any such *deduced* siting criteria would not carry anything like the same force of law as the California criteria in arriving at future siting decisions; the best claim that could be made with them would be that they had set a *precedent.* Hierarchical regimes like precedent, which gives them a non-negotiable frame with, at the same time, considerable space for manoeuver.[20] More egalitarian regimes are suspicious of the inequitable trade-offs between the hierarchical and entrepreneurial biases that might go on within this "manoeuver space" and prefer instead to have everything cut-and-dried and written down in black and white.

The same gross anomalies that emerge when the California siting criteria are applied to the British case also emerge when they are applied to the Dutch and the German cases — both seriously contravene the California population density requirements. On the other hand, although there are differences, the same gross anomalies do not emerge when the British, the Dutch and the German siting decisions are superimposed on one another. All three sites are fairly close to populated areas and the approaches to all three involve quite complex navigation through restricted and much-used channels. The geographical argument that sees these similarities and differences as being dictated by the difference between a small and constricted Europe and a spacious and open California certainly does not hold up in the case of Scotland; nor does it square at all well with the Dutch rejection of the artificial island option that would actually have satisfied the California criteria.

No, we must seriously consider that, like the Duponts placing their family mansion right in the center of their black powder works, these European countries actually prefer that the risks they take for the benefit of the whole be absorbed by that whole. Rather than pursue a goal of zero risk, justified by the sectarian arguments in favor of equity, they strive to institute a judicious measure of *noblesse oblige* — a hierarchic code in which the proper relationships of the parts are given expression through the sacrifices each must be prepared to make to the whole — which they justify by the convergence of two arguments: the entrepreneurial argument in favor of economic efficiency; and the hierarchical argument in favor of logical, orderly, and visible (but not necessarily participatory) procedures.

A growing concern that does not grow

One prescription that is advanced in the FRG case study is that there should be more public participation and less secrecy in the decision process in order that "the growing concern about risk and the negative effects of technological development" may be more adequately taken into account. The argument is that the present representative system, by excluding public debate on safety questions makes itself insensitive to these concerns. Yet one of the most interesting features of this case study is that these "growing concerns" are very little in evidence. "The opposition", we are told, "was not very strong in Wilhelmshaven"; there was "no participation by national or regional environmentalist groups or movements", and such opposition as there was remained focused on local issues.

The growing concern about risk and the negative effects of technological development, it turns out, has its origin in nuclear power, and the fact that it does *not* spill over into LEG should cause us to question this common generalized assumption. Perhaps the generalization is the wrong way round? Perhaps LEG, with its scant public participation, is the German norm and nuclear power the German exception? Perhaps it is not a growing concern about risk that creates a need for public participation but, rather, more public participation that creates the growing concern?

In the case of nuclear power, it is the transition from federal policy to *Länder* implementation that has provided the opportunity for increased public participation. At the *Länder* level local citizens' groups (the *Bürgerinitiativen*) are guaranteed access to the otherwise rather inaccessible process (by a constitutional feature that is somewhat at odds with the German way of doing things, which was inserted by the Allies in the hope that it would provide some sort of check on the development of extreme state authoritarianism). These opportunities for public participation do not arise in the case of other technologies, such as LEG, because the nature of the federal/*Länder* transfer is not the same.[21] But why should public participation result in an increased concern over risk and the negative aspects of technology in general? Cultural theory provides a plausible answer.

If the local citizens' groups are sectarian (and, since they are (*a*) groups and (*b*) external to the hierarchical system, this would appear very likely) then they will follow an appropriately truculent strategy that rejects compromise, pulls back from negotiation, suspects expertise, sets an absolute priority on equity, and sees arguments in favor of efficiency as nothing short of conspiratorial. From this perspective, risk (involuntary, catastrophic, and irreversible) is for distancing oneself from the two pillars of the Establishment — the entrepreneurs and the hierarchists. Small wonder, then, that increased sectist participation in the decision-making process results in a growing concern about risk and the negative effects of technological development!

The cultural approach suggests that increased public participation may well be an appropriate prescription within an adversary culture, but that it is probably inappropriate within a consensus culture (unless, of course, your goal is to hasten the demise of the deferential regime that accompanies such a culture). But this is not to say that there should be no response to those concerns about risks (and technology generally) that gain salience in the sectist bias, only that they should be responded to in the appropriate manner — sympathetically but at arm's length. This, of course, is a brash and overconfident statement of how the cultural theory can generate prescriptions that are sensitive to the notion of appropriateness. It is intended more as a corrective to the sorts of prescriptions that single-problem approaches generate than as an exemplar of prescription making in the plural mode.

The Dutch anomalies

(1) Central government assigned itself a central position yet was confronted by new developments and dynamics apparently outside its direct control — Gasunie's initiative in approaching the Groningen local authorities, for instance.

(2) Ultimate responsibility for a successful outcome rested with the cabinet. "Success", in such a setting, is achieved by mustering convincing justifications for the decision, yet the cabinet chose not to make any use of the one absolutely marvelous justification that was lying there just begging to be used — that in terms of the accident risks at the two sites. The Dutch cabinet did not endorse the Groningen local authorities' view that the Eemshaven risks were acceptable, nor did it involve itself with the opponents of the Maasvlakte site who claimed that the risks there were unacceptably high. But, though it rejected this justificatory basis, the cabinet, in choosing Eemshaven, parted company with many of its official advisory bodies (which were concerned with national rather than local decision making and which favored Maasvlakte on energy policy and economic grounds). Then, having spurned one excellent justification and diverged from its advisors on two other justificatory arguments because they pointed in the other direction, it pinned everything on the wishy-washy socioeconomic argument. Small wonder, given the fragile and fluid coalitions engendered by the Dutch system of proportional representation, that the decision on LNG become one with the very survival of the government.[22]

The first anomaly (an agency claiming overall control and then being continually surprised by its lack of control) is a familiar enough phenomenon that may readily be understood in terms of the asymmetry of an arena in which the initiative lies with the proposer. Surprise, after all, is one of the Principles of War and it is not easy for a sedentary defender to do anything very surprising.

Yet, when we look more closely, we find that Gasunie is not exactly one of the savage beasts of capitalism; it is a semi-state-owned company that has been given a mandate by the government to import LNG within the context of a clearly stated and agreed energy policy. The trade-offs and mutual accommodations between the entrepreneurial and hierarchical biases are here so developed that the whole process is more like a cozy little war game than the real thing. The asymmetry of the arena is so eroded, and the two actors within it have become so complacently reconciled, that it is difficult to see how either could do anything much to surprise the other. Perhaps the real source of the surprise lies *outside* the arena? Perhaps it was a sectarian assault on these complacent accommodations that triggered the surprising events?

Look again, and we see that what prodded Gasunie into independent action was the local concern about the safety of an LNG terminal at

Maasvlakte that might have made official approval increasingly difficult to obtain within the time available. This would suggest that the real source of surprise had to do with an unanticipated shift in cultural bias (towards sectism) as we go from the national to the local government level.

Occam's razor (the Principle of Parsimony) comes to the support of the cultural theory at this point because this explanation, it turns out, also resolves the much more serious problem that is presented by the second anomaly — that the Netherlands case study lies beyond the scope of the framework that is designed to explain it.

> . . . the outcome . . . cannot be interpreted solely in terms of the different official party perspectives and dimensional party interactions, as represented in the MAMP model. In particular the clear divergence between the cabinet's final view and many of its advisory bodies . . . limits the extent to which the cabinet's final decision can be understood in terms of official advice submitted to it.[23]

If the official framework cannot explain what happened, perhaps the unofficial one can. But, to get at the unofficial framework, we have to go right down into the social dynamics that generate and sustain the various objector groups — particularly those at the local level in Rotterdam, Rijnmond, and Groningen — but the Dutch case study, alas, does not provide sufficient information on these groups. But we do have information on these sorts of groups in the British context, and so I will make an oblique approach by first providing a cultural sketch of the British case study and then using it to present a plausible resolution of this serious Dutch anomaly.

A cultural approach to central and local government

Local government, as its name implies, is nested inside something else: national or central government. Of its nature it is hierarchical and, in its tug-o-war with the other pillar of the Establishment — the entrepreneurs — its hierarchical physique is its strength. But what if the sectists enter into the contest as well? The whole thing is thrown off balance. The sectists, above all, are *pro-local* and they demand that local government becomes more local. "Decentralize in the name of Small is Beautiful", and "Redistribute in the name of Equity", they demand. But decentralization requires less hierarchically organized intervention, whilst redistribution requires more. There's the sectist rub!

Local government, by virtue of the fact that it is local, is much more prone to sectist demands than is central government. But, before there can be sectist demands, there have to be some sectists there to make them and this means that the potential cultural mismatch between central and local government is only activated when the sectarian demands are effectively present at the local level. In Scotland, for instance, the local objector group, the Aberdour and Dalgety Bay Joint Action Group (ADBJAG), operated along Establishment lines. They were at home in the setting of a public inquiry, they understood and played by the rules, they respected science and expertise, they sorted out their differences with other objecting groups (the Conservation Society, for instance) beforehand so as to present a consistent and unified front in the debate, and they even sat down to dinner with the enemy. They were respectable rebels.[24] Like nimby (not in my backyard) groups elsewhere in Britain their membership is fairly conservative, fairly elderly, and strongly middle class, and this leads one to suspect that perhaps not all nimbies are equal; that a nimby mobilized from a social fabric richly embroidered with articulate professionals (Aberdour and Dalgety Bay, for example) is likely to be much more formidable than one that emerges from a blue-collar fabric (Canvey Island, for example) and that there may be some threadbare social fabrics so lacking in manipulative and information-handling ability as to be unable to mobilize any sort of nimby at all (Cowdenbeath, for example[25]). The result is that planning proposals end up tripping daintily through the nimbies — giving the formidable ones a wide berth, avoiding the weaker ones where possible and, other things being equal, stepping where there aren't any nimbies at all. As any civil engineer experienced in motorway construction will tell you, the best soil conditions are always to be found in working-class areas.

In the Braefoot Bay case, other things were not equal. There was nowhere else to step (apart from Peterhead, which had already been tried, and some much more expensive alternatives to the north and west), Mortimer's Deep was just about the only economical place capable of accommodating Shell/Esso's colossal foot, and so they really had no option but to meet the nimbies head-on.

But in the Netherlands, the two things were different. First, the available options were such that it was possible for the applications to trip daintily through the nimbies all the way to an unopposed site. Second, the nature of the opposition along the way (I will suggest) was much more sectarian than was the case in Scotland; and it was this sectarian cultural bias that, in giving rise to the mismatch between central and local government levels, produced the Dutch anomalies.

Moral justifications at cross-purposes

One way of understanding local government is to change it. Fortunately, there is plenty of such change about. In Sweden, for instance, the massive reorganization of local government that occurred during the 1970s was justified by a number of claims:

(1) the economies of scale that would be achieved by amalgamation;
(2) the economies of rationalization that would be achieved by amalgamating in such a way that all the new local authorities were of comparable size;
(3) a fairer distribution of services. For instance, smaller authorities

cannot afford even the smallest swimming pool but if they are amalgamated, then everyone will have access to a swimming pool.

This last is a justification in terms of increased equity, and the Swedish assumption has been that increases in equity can be subsumed within the efficiency argument that runs through the first two justifications. Efficiency and equity, so the assumption goes (and it is not just a Swedish assumption), may not be exactly parallel but they run in the same sort of direction. This I will call the *complacent assumption*, and I will question it by proposing a discomforting hypothesis that, far from running in roughly the same direction, equity and efficiency are seriously divergent and perhaps even directly opposed.

Already, some students of local government (those who have tried to evaluate the reorganizations) have suggested that efficiency arguments have a built-in technocratic bias and are anti-democratic, whilst equity arguments have a built-in grassroots bias and are democratic. Translated into cultural theory, what they are saying is that the pursuit of efficiency leads to hierarchy-heightening, whilst the pursuit of equity leads to base-widening.[26] What, then, happens if you insist on both? You increase the height of the governmental pyramid and you increase its base; therefore you increase its volume. If, on the other hand, you cannot increase its volume (because of a world recession, for instance) then you must choose either one or the other. Central government will usually choose efficiency, but local government, in a strongly sectist locale, will be forced to flip the other way. Let me explain this switching mechanism in more detail, for it is the key to the whole argument.

The trade-off between entrepreneurs and hierarchists is shaped by considerations of efficiency and is tempered with two partially contradictory concerns:

(1) the entrepreneurial concern for competition, which is normally justified by the notion of *equality of opportunity* and which is seen as being achieved through the free operation of the market;[27]
(2) the hierarchical concern for order, which is normally justified by the notion of *equality before the law* and which is seen as being achieved by nicely judged interventions aimed at mitigating the worst excesses of the market.

If we call the mediating institutions that are responsible for achieving these trade-offs *government*, then this balancing act between the two powerful cultural biases provides quite a good description of what government, in the context of a consensus culture, has to do in order to moderate each specific debate in such a way as not to erode its general consent. But what happens in an adversary culture where the sects (or, rather, the sect-leaders) are also in on the act?

To the sectarians the trade-off along the *positive diagonal*,[28] between the hierarchists and the entrepreneurs, is an unacceptable and disgracefully inequitable piece of mutual nest-feathering. Their characteristically strident and uncompromising demands are that the trade-off should be radically shifted at right angles to the positive diagonal and strongly in the direction of increased equity. Single issues, irreversible risks, inviolable rights, purity of commitment, refusal to compromise, redistribution acting as a trump card over "expedient" arguments for efficiency, and *equality*, not of opportunity, but *of result* are the hallmarks of the sectarians' position. Their strategy is shaped by a rationality of truculence and their uncompromising moral stance is achieved by a commitment to the *negative diagonal*[29] that connects their social context to those isolated and impotent individuals — the ineffectuals — on whose behalf they credibly claim to speak.[30]

Then, if the sectists are to be persuaded not to withdraw their consent, government will have to strike a much more complex balance. Of course, in a deferential regime, such sectist demands may safely be ignored;[31] the sectist bias is so weak in comparison with the positive diagonal that government would actually stand to lose more consent by moving in the equity direction than it would gain. In Scotland, for instance, only one voice, that of Mr. Jamieson,[32] is raised in the name of sectism. With fine and xenophobic frenzy he castigates both the entrepreneurs and the hierarchists for their willingness to throw away the land that fed a thousand Scots. But the final decision, though it attaches numerous conditions in response to the positive diagonal arguments of the Conservation Society and ADBJAG, pays absolutely no heed to poor Mr. Jamieson. The final decision is a straight balance between the two pillars of the Establishment: the trade-offs run smoothly up and down the positive diagonal.

But, in a truculent regime, the sectist bias is sufficiently strong for a government's departure in the direction of equity to bring with it an increase in consent. Only if it moves *too* far in this direction will the loss in consent from the positive diagonal start to outweigh the increase it is receiving from the negative diagonal. In this way government, if it is to maximize its consent, has to strike a complex three-cornered balance in which the rewards of efficiency that are generated along the positive diagonal have to be diverted at right-angles along the negative diagonal in order to keep all them grassroots a-growin'.

If central government finds itself striking the two-cornered balance, and local government the three-cornered balance, then they will be working against one another. The moral justifications in terms of efficiency that sustain the positive diagonal simply cannot be reconciled with the moral justifications in terms of equity that sustain the negative diagonal. Central government will be acting so as to stabilize a deferential regime whilst local government will be doing all it can to usher in a truculent regime. When all this is referred to the social context diagram (Figure 4) the result is two fundamentally crossed purposes. The efficiency arguments, in providing a negotiating language for the hierarchical and entrepreneurial

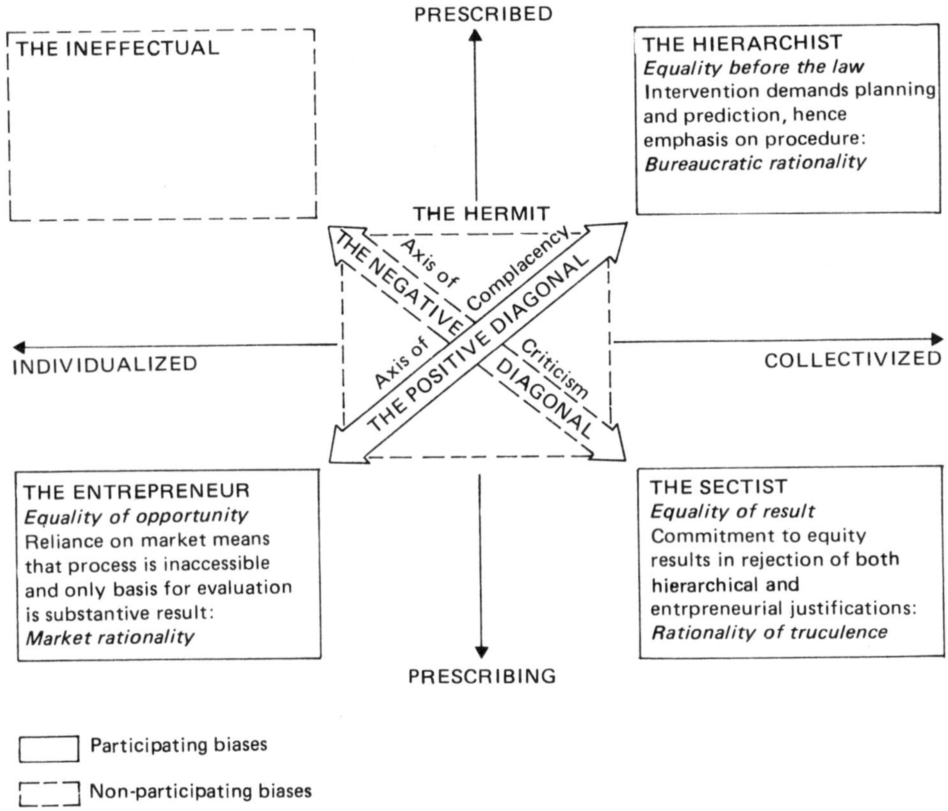

Figure 4. Social contexts, cultural biases, moral justifications, rationalities, and the attraction of opposites.

biases, define an *axis of complacency*; the equity arguments, in providing the sectists with a passive clientele of ineffectuals, define an *axis of criticism*. In this way efficiency and equity are literally at right angles to one another.

The Dutch anomalies revisited

How does this hypothesis — that efficiency and equity are seriously divergent and that central government may go one way and local government the other — explain the Dutch anomalies, and what evidence is there to support it?

First, variations in social fabric as we go from the prosperous and metropolitan Rotterdam area to the less prosperous and more provincial Groningen area would account for a falling off in the strength of nimby opposition and, by the tripping-through-the-nimbies principle, this falling-off would explain why the terminal finally ended up at Eemshaven. Second, if the dominant cultural bias within these opposition groups was a sectist bias, then that would provide the switching mechanism that would put a complacent central government completely out of step with a critical local government (in Rotterdam and Rijnmond). This would explain central government's surprise when it was hit by a sectist assault from outside the complacent arena that it shared with Gasunie, and it would explain the curious pattern of justification subsequently adopted by the cabinet.

In tripping through the nimbies, the final decision passed out of the effective control of a deferential regime and, thereafter, followed the line of least resistance that led it away from the consensus-lacking truculent regimes around Rotterdam and towards the nimby-free (and consensual) Groningen area.[33] So the decision was really determined by the equity-focused criteria of truculent local government regimes, but it had to be justified in terms of the efficiency-focused criteria of a deferential central government regime. In bowing to the strength of the sectist arguments around Rotterdam, central government had to part company with its positive diagonal advisors yet, in justifying that decision, it could not be seen to espouse the sectist risk arguments that, in fact, decided it all. What evidence is there to support this argument?

Clearly, Gasunie found itself dealing increasingly with the local government world. Directly, and indirectly, it received indications that there was a lack of consensus on the acceptability of LNG in the Rotterdam area. But central government kept itself remote from these dealings and the first thing it "knew" about the reintroduction of the Eemshaven site was when it received a request from the Groningen local authorities who had by that time reached a satisfactory accommodation with Gasunie. That this arrangement was satisfactory suggests that, unlike in the Rotterdam area, there was a consensus on LNG in Groningen. So what was responsible for the lack of consensus in the Rotterdam area?

(1) The case study tells us that reports in the media indicated that the main issue was the impact on the local environment and, in particular, the increased danger that LNG would pose to the local population. Lack of consensus on criteria for deciding *factual* risk led to concerns being aired about the *perceived* risk level to the local population.

Orr's table tells us that environmental change and technological accidents are the risks that gain salience in the sectist cultural bias, and the fact that these concerns were pressed with sufficient force to dislodge the complacent axis's determination of factual risk, and to substitute a higher perceived risk, indicates that the sectist bias was present in some force and that it was effective.

(2) A serious threat of delay was presented by "the interest which was shown by some (in particular within the Rijnmond Public Authority) to design a unique 'public participation' program as part of the approval procedure at the local level."[34]

Though the case study does not tell us who those "some" were, we can guess. As local government gets more local, so increased public participation (if it is sectist) erodes the consensus that has been reached at the higher levels (central government and Rotterdam).

(3) The case study has to appeal to the presence of "political pressures" to account for the switch from Maasvlakte to Eemshaven. Groningen, we are told, had got its act together and was able to lobby effectively, whilst in the Rotterdam region the local authorities were divided. In addition, the condition demanded by Rotterdam — that they would not get a nuclear power station as well as an LNG terminal — helped sway the decision in favor of Eemshaven.

But why did Rotterdam impose this condition and why did Groningen not impose it? And why was Rotterdam divided and Groningen together? The cultural hypothesis has the perfect explanation.

The presence of a strong sectarian bias in Rotterdam (and particularly in Rijnmond) would explain the lack of consensus there, and the absence of such a bias in the Groningen area would explain the presence of consensus there. On top of this, the great "risks-for" in the sectarian bias are the involuntary, irreversible, and catastrophic consequences that are perceived to accompany nuclear power. So the presence of a sectarian bias in the Rotterdam area would account for the nuclear plant condition there, and the absence of such a bias in the Groningen area would account for the absence of the condition there.

(4) The cabinet's overriding emphasis on the socioeconomic argument (regional development policy, unemployment, etc.) in justifying its final decision in favor of Eemshaven is readily understood when we see that its official advisors (on the grounds of economic costs and energy policy) favored Maasvlakte. But it does not help us to understand why it chose Eemshaven in the face of these two efficiency arguments; nor does it explain why, having made the decision, it did not draw on the readily available safety argument as a powerful justification for that decision.

The cultural hypothesis (and in particular the switch from deference to truculence as we go from the central to the local government level) explains why central government, faced with a sectist *fait accompli* at the local level, had no option but to decide in favor of Eemshaven. At the same time, it explains why the cabinet, as an institution committed to the stabilization of a deferential regime, could not possibly be seen to use the sectarian safety arguments. They are the arguments appropriate to the stabilization of a truculent regime and, as such, would justify the capitulation of a deferential regime.

An alternative interpretation that is still consistent with the cultural hypothesis (but which places less emphasis on the switch from hierarchical to sectarian influence as we go from central to local government) sees a culturally rather inconsistent cabinet stumbling, as best it can, from one crisis to the next. In the sorts of coalition systems that follow from proportional representation, sectism is able to penetrate government much more strongly than in the "two-party, first past the post" system. All sorts of people (our case study writer, for instance) can gain access to the upper reaches of the Dutch government and civil service with an ease and legitimacy that would never be tolerated in Britain or the FRG. The result is a lack of cultural consistency and a rather high degree of arbitrariness, with power receding to more private (and more culturally consistent) cliques, and with the cabinet's capacity shrinking to piecemeal bilateral agreements and deals (like Gasunie's). Within this interpretation, the cabinet's shunning of the justification in terms of accident risk is seen as not so much an assertion of a strongly hierarchical cultural bias, but rather as a stopgap measure aimed at avoiding setting a precedent that might reduce its scope and manoeuverability, even further.

Perhaps, pushing the hypothesis to its speculative limit, these two interpretations are best seen not as alternatives but, rather, as fairly distinct phases within a single continuous process of transformation. First, the sectarian influence increases at the local level and central government (still hierarchical) fights back; then sectarianism, diffusing upwards, pluralizes even the cabinet, breaking down its cultural consistency and rendering its decision making increasingly arbitrary and hand-to-mouth; finally, power migrates to more private and culturally more consistent cliques that can then use the public organs of government as little more than legitimizing rubber stamps for their multifarious deals.

TECHNOLOGIES AS CULTURAL PRODUCTS

Coming to LEG risk from nuclear power and from smoking and health, the most interesting thing about this debate is that it is so boring. Shell and Esso, admittedly, are no *more* boring than British Nuclear Fuels or Philip Morris;[35] the tedium has to do with the almost complete absence of variety among the groups and individuals, who, though not the instigators of the proposals, nevertheless feel sufficiently moved by them to speak out — either for or against. In the Shell/Esso corner we find no counterpart of SE_2 (Scientists and Engineers for Secure Energy) and its awesome spokesperson, Ed Teller — "the Father of the H-bomb"; in the other corner, instead of a cacophonous assortment of anti-groups with wondrously assorted acronyms like GASP (Group Against Smokers' Pollution) or SCRAM (Scottish Campaign to Resist the Atomic Menace), there is just one quite well behaved nimby — the Aberdour and Dalgety Bay Joint Action Group — with an instantly forgettable set of initials that cannot even be pronounced.

What makes this absence of excitement so exciting is that it runs counter to the current conventional wisdom that sees everywhere a growing concern about the negative aspects of technology and that then seeks to explain this general phenomenon in terms of a widespread transition from industrial to post-industrial society. According to this prevalent hypothesis, once a certain level of affluence has been reached, there comes an inevitable shift from material to spiritual values. This shift, by diminishing the positive (but material) advantages of modern technology and by inflating its negative (and spiritually damaging) disadvantages, radically alters our whole attitude to the innovative flow that has been so central a feature of Western society since the start of the Industrial Revolution.[36] But this explanation is of the "single problem—single answer" type and, as such, deserves to be viewed with suspicion.

Why, if there is everywhere a growing disenchantment with technology, was there no participation of national or regional environmentalist groups or movements at Wilhelmshaven? In Scotland, why were the Friends of the Earth not there? Where was the Oxford Political Ecology Group? These are not idle questions. The German Greens have been vigorously active in the nuclear debate and at Frankfurt Airport; the British Friends of the Earth and the Oxford Political Ecology Group were there in force at the Windscale Inquiry (as were SCRAM and other anti-nuclear power groups), and both (unlike SCRAM) had legitimate reasons for wanting to be at the Mossmorran—Braefoot Bay inquiry as well. The British FoE, it turns out, would have liked to have been there but they had devoted all their resources (a total of £40 000) to the Windscale Inquiry which, fortunately for Shell/Esso, was taking place at virtually the same time. But why did the FoE choose to allocate their resources so disproportionately between these two technologies? If LEG is likely to kill as many people as uranium oxide reprocessing (and, unlike reprocessing, it has already killed a considerable number), why does it have such a low priority and why, conversely, does nuclear power have such a high priority? What rationale lies behind such seemingly irrational behavior?

Since it lacks the requisite variety, the post-industrial society theory is powerless when faced with such marked variations in response as we go from one technology to another. But cultural theory can advance the following hypothesis in terms of the hidden agendas (the "risks-for") that are built into each of the cultural biases. In each cultural bias the risks are selected according to how effective a lever they provide for the advancement of the hidden agenda — the desired pattern of social relations — appropriate to the stabilization of that cultural bias. These "risks-for" are to be understood in terms of the internally generated requirements for stabilizing the various organizational types that underlie and, at the same time, are sustained by the various cultural biases.[37] What, then are the organizational requirements of the bounded egalitarian group — the sect — and what kinds of risks are best suited, culturally, to its organizational purposes?

The sect member's overriding commitment to equality prevents the development of any of those internal differentiations that characterize the hierarchically nested group. The sect's structure, therefore, has to be concentrated at the edge of the group, at the point where it cuts itself off from the nasty, inequitable outside world. To justify, and to keep on justifying, the constant vigilance needed to maintain and defend this crucial boundary you need a suitably powerful, greedy, and villainous besieging force. By their risks ye shall know them:[38] massive concentrations of energy high-handedly set down on the doorsteps of those least able to resist; the impoverishment (perhaps, even, the extinction) of future generations for the short-term benefit of a power-crazed elite; and the insidious destruction of those who dare to speak up against it all.

For a bounded egalitarian group, according to the cultural hypothesis, an issue is likely to have a high priority:

(1) If it entails the physical concentration of energy and the social concentration of its control. Concentrated energy is the life-blood of the corporate state, and the specialized institutions that channel it are the elite and vital organs of its circulatory system. *Energy*, as Aaron Wildavsky has remarked, *is to sectarians as red meat is to vegetarians.*

(2) If it involves risks that extend to the long term; better still, if some of the risks are restricted to the long term; best of all if they are also irreversible, involuntary, and catastrophic. The entrepreneur, notoriously, will sell his grandmother; the hierarchist will sacrifice the part to the whole; both, in their different ways, discount the future. Only the sectist cares ceaselessly about the meek, the helpless, and the innocent. And what could be more perfectly inequitable than the imposition, as a result of decisions taken today, of appalling and incalculable risks on future generations? Identify and broadcast the long-term, involuntary, irreversible, and potentially catastrophic consequences of some new technology and you can make credible in the here-and-now the most dreadful, the most despicable, and the most indefensible of all the moral outrages that are perpetrated by the positive diagonal: *the massacre of the unborn innocents.*

(3) If it brings with it risks that threaten the sinister and unseen penetration of the body. One apparent paradox of the sect-like organization is that its puritan bias tends to restrict the supply of media for (among other things) the expression of the social concerns that will sustain its existence; often the supply is restricted to just the physical body as a metaphor for the social body.[39] But, since the human body, with its fragile skin and its tempting orifices, provides an apt and powerful natural symbol for the soft, vulnerable sect forever threatened by a nasty, devious, and predatory "them", this is no great disadvantage. The fear is not so much of straightforward rape as of the fatal corruption produced by invisible penetration or by agents that, though

> visible, are not what they appear to be (witchcraft and conspiracy fears). Concede the validity, the power, and the organizational good sense of this natural symbol and many a seemingly emotional and irrational sectarian response clicks into place: *corporate capitalism causes cancer.*[40]

Nuclear power scores very highly on all three of these counts. Indeed, if you set out deliberately to design a technology that would provoke the maximum sectist opposition you could scarcely do better. LEG, on the other hand, scores badly. True, it involves a massive concentration of energy, but it is a *clean* concentration. If it is going to kill you it will freeze or fry you instantly, producing some water vapor and a little carbon dioxide that soon disperse leaving no poisonous or persistent residue. The risks in LEG (apart, perhaps, from those involving ignition caused by radio waves) are honest, instantaneous, and visible, and they do not get inside the body. By these cultural criteria, LEG provides little by way of leverage for the advancement of the sectist's hidden agenda.[41]

Now, equipped with this cultural hypothesis, one can say that the FoE's allocation of resources between nuclear power and LEG is perfectly rational. And one can go on to say that the overspill from nuclear to other issues (such as LEG), whilst possible, is not simply mechanical. It is not that there is everywhere a growing disenchantment with technology, nor is it that some "pathfinder" technology has cleared the way for other more fortunate technologies to follow largely unopposed.[42] Rather, containment and overspill have to be *socially achieved*, and the "risks-for" criteria will provide us with a useful basis for assessing which way any particular technology is likely to go.

Of course, the fact that the different cultural biases are in contention means that containment and overspill do not depend on just this sectarian assessment. The technology has always to be assessed in relation to the regime (the particular mix of biases) that is handling it. Although the nuclear case is rather special, and (other things being equal) likely to be contained, it may nevertheless spill over into other technologies because of the strength of the negative diagonal's criticism. Mature and confident establishments know how to bend so as to incorporate serious opposition but the pro-nuclear establishment (in Britain and the FRG, at least) has not bent and, losing its hierarchical confidence, has itself taken on somewhat sectarian characteristics — for instance, by emphasizing the general apocalyptic consequences that will ensue if it does not get its way ("Stone Age? No thanks" stickers in response to the anti's "Nuclear Power? No thanks"). The result of this fluid contention between increasingly effective sects and badly rattled hierarchies has been a waning of compromise and a broadening of what was initially a quite narrow issue, and this generalization has led to the incorporation of other concrete issues

(Frankfurt Airport, for instance) and to an erosion of the legitimacy of many overcomplacent institutions.[43]

CONCLUSION

One major source of this complacency has been *technology assessment* and its offspring, *risk assessment*, which have fastened, with technocratic tenacity, on just the technology itself and not on the relationship between the technology and the temper of the people that the technology is intended to serve. But man cannot live by multi-attribute utility functions alone; and the narrow rationality that speaks of the prevalent misperceptions of lay people, that insists on the explicit trade-off of risks and benefits, and that pretends that the open-systems art of engineering is just a low-status branch of bench science, has been knocked sideways by the unwillingness of those misperceivers to realign their vision with that of the experts, by the refusal of those who hold certain things sacrosanct to exchange them for things they do not hold sacrosanct, and by the distrust that has greeted those who claim to have anticipated and quantified everything that can go wrong.

In this Postscript I have advanced cultural theory as a way of transferring the focus of attention away from the technology itself and onto the relationship between the technology and the regime that is handling it. Unlike the "single problem—single answer" approach that pervades both technology assessment and its radical mate, post-industrial society theory, cultural theory assumes a plurality of biases that are locked in continuous, dynamic, and fluid contention. Material and spiritual values are always with us; it is not their presence or absence that matter, but the manner, the influence, and the stability of their mixes. At the same time, by tracing these biases back to the stabilizing needs of different types of social organization, cultural theory avoids that counterpart to the universalist fallacy — the individualist fallacy — that insists that every individual must be treated as a special case.

That said, I should stress that both the cultural theory and the interpretations of the case studies that have been offered here are not intended to be definitive and final. Rather, they should be seen as exploratory and suggestive; as ways of opening up fresh prospects and of pointing towards the sorts of directions in which it may be worth looking in the future. The essential difference is that the "single problem—single answer" theories assume generality and are then powerless to explain variation, whilst the cultural hypothesis assumes that there will be variation and then looks to particular historical, social, and symbolic conjunctions to explain particular instances of generality when they occur.

NOTES

1. Subject to clearance on seismic risk — a Californian preoccupation that did not enter into the British debate. (Clearance was finally obtained in 1982).

2. Kenneth Boulding *National Defense Through Stable Peace*. Lectures delivered at IIASA, June/July 1981

3. For an elaboration of this argument see my *Rubbish Theory* (Thompson 1979).

4. See Douglas and Wildavsky (1982).

5. The universalistic position leads to the impossible claim of the "cosmic exile"; the totally relativistic position leads to the serious consideration of all "so-called possible entities". For a philosophical approach, see Quine (1953).

6. Douglas (1978b).

7. Orr (1977).

8. These perspectives clearly have much in common with those discerned, independently, by Harold Linstone and others (Linstone *et al.* 1981, see also Thompson 1982).

9. This normative/interpretive trap has long been a topic of concern in anthropology and it has resulted in an important distinction between two sorts of approaches — the *etic* approach, in which the analyst defines the relevant categories and then fits his exotic society into that framework in much the same way that poor Procrustes was fitted to his couch; and the *emic* approach, in which the analyst tries to discover the categories of the people he is studying and then interprets that society in terms of those categories. The etic approach is likely to be strongly normative and highly ethnocentric and it results in what is called *the anthropologist's model*; the emic approach, bending over backwards to avoid these twin evils, results in what is called the *home-made model*. Of course, once a number of these home-made models have been constructed, it is perfectly legitimate (from the emic point of view) then to put them beside one another and to try to construct a meta-model, as it were, to explain their variance. From the emic point of view the anthropologist's model is valid *only* at this meta-level (where, of course, it will have to compete with the rival models of other anthropologists). That said, it only remains for me to add that I am an anthropologist and that nowadays no self-respecting anthropologist could possibly bring himself to take an etic approach. (Leastways, not knowingly! Just

because you are striving to compensate for bias it does not follow that you are entirely free of it.) The problem for me, as a member of an interdisciplinary team, is to find an emic way of handling these case studies; the problem for the team, as a whole, is to find some non-arbitrary basis on which to compare them. I will argue that the solution to the first problem solves the second problem as well.

10. Often the decision is already taken for him by the unquestioned assumptions that are built into the social context of the analysis. Technical fixers, for instance, have an idea of nature as essentially fixable and, for them, all risks lie within that nature.

11. Also, as Brian Wynne has pointed out to me, MAMP should be criticized for lobotomizing the interest groups — for defining *them* according to *its* framework, rather than getting a feel of their concerns and interests "in the round" and then fitting the LEG siting issue into that social and plural frame.

12. A parallel with organization theory may help to illuminate this crucial distinction. A hierarchical organization will, by virtue of its structure, define for itself a *manifest goal*, but only in the smallest of such organizations will this be the *only* goal that is being pursued. As different information cultures crystalize out in different sectors and levels within the organization, so they create for themselves *latent* goals that are likely to be in conflict both with one another and with the manifest goal. This means that an analysis in terms of just the manifest goal will be hopelessly inadequate since the behavior of the organization will make sense only in terms of the complex patterns of contentions – both those between the protagonists of the manifest and latent goals and those between the protagonists of the various latent goals.

13. Lévi-Strauss (1966).

14. But the tactical uses of references to resources may differ from these underlying strategic categories. Entrepreneurs for example, will say there is resource depletion in one sector (hydrocarbons or, even, uranium) to justify business-as-usual and resource abundance in another (fast breeder reactors). For a discussion of the relationship between strategic and tactical uses see my *Among the Energy Tribes* (Thompson 1982).

15. I should stress that here I am following the definition of an organization as a *conceptual scheme*. I do not wish to suggest that the concrete reality — the process of social life — crystalizes out so neatly. In general, this process is sufficiently complex and messy for anyone involved in it to be able to conceive it, and render a plausible account of it, in one of these three ways. The patternings and transformations

of this "concrete reality" are to be understood as the resultant of these contradictory conceptual schemes as they are acted upon by those who variously hold to them.

16. Ineffectuals form, as it were, a reservoir of social capital that each of the other social contexts (except the hermit) tries to appropriate, and then transmute, with the help of its cultural bias. Whilst the sectarians need them for a moral bludgeon, the entrepreneurs need them as both a pool of labor and a placid source of consumption, and the hierarchists need them beneath them, as wayward children or as cannon-fodder, according to circumstances.

17. Indeed, in the LEG case, need went on and on being debated even after it was supposed to have been established that there was a need for the terminal — a nice example of visible and hidden agendas in contention.

18. Within a deferential populace sects can gain little standing, because the hierarchies have managed to achieve and sustain an impressive maturity. But, within a truculent populace, sects do have standing and the result of their continual whistle-blowing is the imposition of a permanent immaturity on such hierarchies (government bureaucracies, for instance) as do form.

19. Indeed, two countries that are reputed to be the most hierarchical of all — France and Japan — were actually dropped from our study in its early stages because their "universes" were *so* light.

20. But this is not to say that they like *all* precedents. We will presently look at the remarkable contortions that the Dutch cabinet had to perform in order to avoid setting a precedent that would have reduced its manoeuver space.

21. And, probably, for some other reasons as well. These have to do with the social achievement of overspill and containment and are discussed later.

22. The decision on LNG *could* have endangered the government by precipitating a no-confidence vote in parliament. Parliamentary sources, however, play down the likelihood of this possibility.

23. The exact wording is taken from the full IIASA case study (Schwarz 1982).

24. King and Nugent (1979).

25. Cowdenbeath is a depressed one-time mining community with a very high level of unemployment. It is near to both Mossmorran and Braefoot Bay and its inhabitants (when asked by television interviewers) are almost unanimous in their support for the project which

they believe (wrongly in the opinion of many experts) will bring them the employment they crave. They are dismissive of the risks entailed in LEG technology, not so much because they have clear perceptions of these risks as being lower than their questioners suggest, but because they need work and are prepared to accept a high level of risk to obtain it. In the happy days when they had work it was in the now defunct coal mines and they claim, with some pride, to be used to living with a high level of risk. So their perceptions of risk are consistent with the predictions for the ineffectuals' social context, as is their evident lack of control over their destiny. They, clearly, have every bit as much incentive to form themselves into a pro group as the articulate professionals of Braefoot Bay have to form themselves into an anti group, yet the pro group does not form whilst the anti group does. At the Edinburgh Airports Inquiry in the early 1970s, Newhouse — a socially similar community to Cowdenbeath — displayed exactly the same lack of involvement.

26. Whilst the pursuit of efficiency leads to hierarchy-heightening, this latter does not always lead to increased efficiency. The relationship is curvilinear — hierarchy judiciously introduced in small doses often does increase efficiency but, beyond a certain level, hierarchy-heightening (empire-building) results in inefficiency. The same is probably also true of the pursuit of equity.

27. What they actually do may not always accord with this moral justification. They may be happy to have no competitors (monopoly) or high entry barriers (oligopoly) so long as they are winning!

28. So called because it links the two power-wielding contexts.

29. So called because it is sustained by its opposition to the positive diagonal.

30. This is not to say that those in other social contexts are not concerned for the ineffectual; only that the sectarian strategy "singles them out", whereas the others try to "fit them in" (the hierarchist strategy), or "melt them down" (the entrepreneurial strategy — "what's good for General Motors is good for the United States, including the weakest, you understand!)

31. Indeed, it is the ignoring of sectist demands (other than in an indirect, arms-length, way) that stabilizes a deferential regime. I do not wish to imply that regimes and patterns of bias are clearly separable. Rather, both are to be seen as the mutually reinforcing conditions for certain recurrent regularities to be possible within a continuous flux.

32. Mr. Jamieson does not receive a mention in the British case study, but, as well as being a member of the ADBJAG, he appeared at the

public inquiry as an individual objector. He also figured prominently in the two television documentaries.

33. There were *some* nimbies there but they reluctantly conceded that, in a straight contest with Rotterdam, the Eemshaven site was considerably safer (mainly because of its lower population density) so they didn't really have to refer to TNO's risk analysis which also confirmed this difference.

34. The exact wording is taken from the full IIASA case study (Schwarz 1982).

35. See my "Fission and fusion in nuclear society" (Thompson 1980c).

36. Inglehart (1977); OECD/Interfutures (1979).

37. It should be stressed that neither the cultural bias nor organizational type is to be seen as logically or causally prior to the other; rather, their mutuality is the necessary condition for their continued existence.

38. With apologies to St Matthew VII:20.

39. See Owen (1982)

40. Another Wildavsky aphorism.

41. The very title of the FoE book — *Frozen Fire* (Davis 1979) accurately reflects the immediacy and cleanliness of the risks that accompany this technology. The fact that both the energy spokesman of the British FoE and the author herself declined invitations (rather lukewarm, admittedly) to the IIASA LEG Task Force Meeting lends support to the interpretation that they have tried to do their best with this technology but have found it to be sadly lacking in delayed and insidious effects.

42. See Häfele (1974).

43. Ronge (1982).

Consultants and Reviewers

The authors gratefully acknowledge the valuable assistance of the following individuals in the preparation of this book.

Bruce Ackerman, Columbia University, New York City, USA
William Ahern, California Public Utilities Commission, San Francisco, California, USA
Brian Arthur, Stanford University, Stanford, California, USA
Jesse Ausubel, US National Academy of Sciences, Washington, DC, USA
Robert Arvedlund, Environmental Evaluation Branch, Federal Energy Regulatory Commission, Washington, DC, USA
Ernst Baumann, Swiss Reinsurance Company, Zürich, Switzerland
David Bell, Harvard Business School, Boston, Mass., USA
Theodore C. Bergstrom, Department of Economics, University of Michigan, Ann Arbor, MI, USA
Arthur Boni, Science Applications, Inc., La Jolla, California, USA
David Bull, Scheggenstrasse 24, 8620 Wetzikon ZH, Switzerland
Ian Burton, Institute for Environmental Studies, University of Toronto, Canada
Niall Campbell, Scottish Development Department, Edinburgh, Scotland, UK
Richard Caputo, Jet Propulsion Laboratory, Pasadena, California, USA
Kimball C. Chen, Energy Transportation Group Inc., New York, USA
Robert Chen, University of North Carolina, Chapel Hill, NC, USA
C.D.J. Cieraad, Technology for Society Division, TNO, Apeldoorn, The Netherlands
L.A. Clarenberg, Rijnmond Public Authority, Rotterdam, The Netherlands
Jobst Conrad, International Institute for Environment and Society, Science Center, Berlin, FRG
Robert W. Craig, The Keystone Center, Keystone, Colorado, USA
Norbert Dall, Alliance for Coastal Management, California, USA
Robin Dennis, Environmental and Societal Impacts Group, National

Center for Atmospheric Research, Boulder, Colorado, USA
Randolph W. Deutsch, California Public Utilities Commission, San Francisco, California, USA
James Dooley, University of Toronto, Toronto, Canada
Elisabeth Drake, Arthur D. Little, Inc., Cambridge, Mass., USA
George Eads, School of Public Affairs, University of Maryland, College Park, Maryland, USA
Ward Edwards, Social Science Research Institute, University of Southern California, Los Angeles, California, USA
Tibor Englander, Institute of Psychology, Hungarian Academy of Sciences, Szondy u83--85, Budapest, Hungary
William Fairley, Analysis and Inference, Inc., Boston, Mass., USA
E.T. Ferguson, The Foundation IIASA — Netherlands, The Hague, Netherlands
Jörg Finsinger, International Institute of Management, Science Center, Berlin, FRG
Baruch Fischhoff, Decision Research, Eugene, Oregon, USA
Orio Giarini, Association Internationale pour l'Etude de l'Economie de l'Assurance, Geneva, Switzerland
Meredith Golden, University of North Carolina, Chapel Hill, NC, USA
Walter A. Hahn, Congressional Research Service, The Library of Congress, Washington, DC, USA
C. Robert Hall, National Association of Independent Insurers, Des Plaines, Illinois, USA
C.S. Holling, IIASA, Laxenburg, Austria
David Huard, Federal Energy Regulatory Commission, Washington, DC, USA
Patrick Humphreys, Decision Analysis Unit, London School of Economics, London, UK
Robert Inhaber, Oak Ridge National Laboratory, Oak Ridge, Tennessee, USA
dr. ir. J.L.A. Jansen, Velp, The Netherlands
Helmut Jungermann, International Institute for Environment and Society, Science Center, Berlin, FRG
Roger Kasperson, Clark University, Worcester, Mass., USA
Ralph Keeney, Woodward-Clyde Consultants, San Francisco, California, USA
Paul Kleindorfer, Decision Sciences Department, The Wharton School, Philadelphia, Pennsylvania, USA
Jerry Kopecek, Energy Resources, Inc., La Jolla, California, USA
Oleg Larichev, USSR Academy of Sciences, Moscow, USSR
Peter Louckes, Cornell University, Ithaca, NY, USA
Charles E. Lindblom, Yale University, New Haven, Connecticut, USA
Harold A. Linstone, Futures Research Institute, Portland State University, Portland, Oregon, USA

Keith McKinney, Western LNG Terminal Associates, Los Angeles, California, USA
Giandomenico Majone, IIASA, Laxenburg, Austria
P.D. Mehta, Aberdour and Dalgety Bay Joint Action Group, Aberdour, Scotland, UK
Louis Miller, Decision Sciences Department, The Wharton School, Philadelphia, Pennylvania, USA
Gavin H. Mooney, Health Economics Research Unit, Universiy of Aberdeen, Scotland
J.S. Nalevanko, Research and Special Programs Administration, US Department of Transportation, Washington, DC, USA
Dorothy Nelkin, Cornell University, Ithaca, NY, USA
Frederick Niehaus, International Atomic Energy Agency, Vienna, Austria
H.J. Nikodem, Battelle Institut, Frankfurt, FRG
Michael O'Hare, John F. Kennedy School of Government, Harvard University, Cambridge, Mass., USA
David Okrent, Chemical, Nuclear, and Thermal Engineering Department, School of Engineering and Applied Science, University of California, Los Angeles, California, USA
Timothy O'Riordan, School of Environmental Sciences, University of East Anglia, Norwich, UK
Yuri Oseredko, All-Union Institute for the Gas Industry, Kiev, USSR
H.J. Otway, Commission of the European Communities, ISPRA, Italy
Elisabeth Paté-Cornell, Stanford University, California, USA
Mark Pauly, University of Pennsylvania, Philadelphia, Pennsylvania, USA
Lloyd Philipson, J.H. Wiggins Co., Redondo Beach, California, USA
drs A.P.J. Planken, Moerkapelle, The Netherlands
Charles R. Plott, Division of the Humanities and Social Sciences, California Institute of Technology, California, USA
Howard Raiffa, Harvard University, Boston, Mass., USA
Jerry Ravetz, Division of History and Philosophy of Science, The University of Leeds, Leeds, UK
Lucas Reijnders, Stichting Natuur en Miliëu, Utrecht, The Netherlands
mr. A.A.T. van Rhijn, Commission of the European Communities, Brussels, Belgium
Volker Ronge, Infratest Sozialforschung, München, FRG
Susan Rose-Ackerman, Columbia University, New York City, USA
William Rowe, American University, Washington, DC, USA
Werner Salz, Bundesministerium für Forschung und Technologie, Bonn, FRG
Paul Schoemaker, University of Chicago, Chicago, Illinois, USA
Behrend Schütz, Gesellschaft für Reaktorsicherheit, Garching, FRG
J.J. Schwarz, Centre for Technology and Policy Studies, TNO, Apeldoorn, The Netherlands
Craig Sinclair, Scientific Affairs Division, NATO, B110 Brussels, Belgium

Dr Schwerter-Strumpf, Bezirksregierung Weser-Ems, Oldenburg, FRG
David Snowball, School of Geography, University of Leeds, Leeds, UK
Nicholas Sonntag, Environmental and Social Systems Analysts Ltd, Vancouver, BC, Canada
Chauncey Starr, Electric Power Research Institute, Palo Alto, California, USA
Michael Stoto, John F. Kennedy School of Government, Harvard University, Cambridge, Mass., USA
Peter Temin, Massachusetts Institute of Technology, Cambridge, Mass., USA
Richard Thaler, Cornell University, Ithaca, New York, USA
dr. W. Turkenburg, Physics and Society, State University of Utrecht, The Netherlands
James Vaupel, Duke University, Durham, North Carolina, USA
Ilan Vertinsky, University of British Columbia, Vancouver, BC, Canada
Carl C. Walters, Institute of Animal Resource Ecology, University of British Columbia, Vancouver, BC, Canada
Gilbert White, University of Colorado, Boulder, CO, USA
Aaron Wildavsky, Survey Research Center, University of California, Berkeley, California, USA
Thomas Willoughby, Assembly Committee on Resources, Land Use, and Energy, Sacramento, California, USA
Alvin Weinberg, Oak Ridge National Laboratory, Oak Ridge, Tennessee, USA
Detlof von Winterfeldt, Social Science Research Institute, University Park, Los Angeles, California, USA
Brian Wynne, University of Lancaster, Bailrigg, Lancaster, UK
Peyton Young, University of Maryland, College Park, Maryland, USA
Richard Zeckhauser, John F. Kennedy School of Government, Harvard University, Cambridge, Mass., USA
Paul Zigman, John F. Kennedy School of Government, Harvard University, Cambridge, Mass., USA

The authors of the case studies gratefully acknowledge the following organizations for their cooperation and information during the course of the research.

Federal Republic of Germany
Bezirksregierung Weser-Ems, Oldenburg
Bundesminister für Verkehr, Bonn
Bundesminister für Wirtschaft, Bonn
Deutsche Flüssigerdgas Terminal GmbH, Wilhelmshaven
Gemeinde Wangerland
Initiativausschuss Hooksieler Vereine, Wangerland

Landkreis Friesland, Jever
Niedersächsischer Minister für Wirtshaft und Verkehr, Hannover
Niedersächsischer Sozialminister, Hannover
Ruhrgas AG, Essen
Ruhrgas LNG Flüssigerdgas Service GmbH, Essen
Stadt Wilhelmshaven: Stadtverwaltung, Wilhelmshaven
Stadt Wilhelmshaven: Stadtrat, Wilhelmshaven
Wasser- und Schiffahrtsdirektion Nordwest, Aurich

The Netherlands
Havenbedrijf der Gemeente Rotterdam
Havenschap Delfzijl
Ministerie van Binnenlandse Zaken
Ministerie van Economische Zaken
Ministerie van Onderwijs en Wetenschappen (Wetenschapsbeleid)
Ministerie van Sociale Zaken
Ministerie van Verkeer en Waterstaat (ICONA)
Ministerie van Volksgezondheid en Miliëuhygiene
Ministerie van Volkshuisvesting en Ruimtelijke Ordening
NV Nederlandse Gasunie
Openbaar Lichaam Rijnmond
Provinciale Griffie van Groningen
Provinciale Griffie van Zuid-Holland
Stichting Natuur en Miliëu
TNO, Nederlandse Organisatie voor Toegepast-Natuurwetenschappelijk Onderzoek
Werkgroep Noordzee

The United Kingdom
Aberdour and Dalgety Bay Joint Action Group
Department of Energy
Fife Regional Council Planning Officers
Health and Safety Executive
Kirkcaldy and Dunfermline District Planning Officers
Scottish Development Department
Employees of Shell and Esso

The United States
Alliance for Coastal Management
Arthur D. Little, Inc.
Assembly Committee on Resources, Land Use, and Energy
California Public Utilities Commission
Energy Resources, Inc.
Federal Energy Regulatory Commission
Science Applications, Inc.

References

Aberdour and Dalgety Bay Joint Action Group (1979) *Mossmorran–Braefoot Bay: Shipping Hazards* (mimeo) (Aberdour, UK).

Ackerman, B., Rose-Ackerman, S., Sawyer, J., and Henderson, D. (1974) *The Uncertain Search for Environmental Quality* (New York: Free Press).

ADL (1978) *Draft Environmental Impact Report for Proposed Point Conception LNG Project*, prepared for California Public Utilities Commission; and *LNG Safety*, Technical Report 16 of Draft EIR for Point Conception (Cambridge, MA: Arthur D. Little, Inc.).

Ahern, W.R. (1978) *Energy Facilities and the California Coastal Act* (San Francisco: California Coastal Commission)

Ahern, W.R. (1980) "California meets the LNG terminal" *Coastal Zone Management Journal* 7:185–221.

Allison, G.T. (1971) *Essence of Decision* (Boston: Little, Brown and Company).

Allison, G.T., Carnegab, A., Zigman, P., and De Rosa, F. (1981) *Governance of Nuclear Power*, Report to the President's Nuclear Oversight Committee, September.

Arrow, K. (1963) *Social Choice and Individual Values* (New York: Wiley).

Atz, H. (1982) *Siting and Approval Process for an LNG Terminal at Wilhelmshaven: A Case Study in Decision Making Concerning Risk-Prone Facilities in the Federal Republic of Germany* Collaborative Paper CP-82-62 (Laxenburg, Austria: International Institute for Applied Systems Analysis).

BAM (1979) *Sicherheitstechnische Beurteilung einer Transport- und Umschlagbrücke vor dem nördlichen Teil des Voslapper Grodens in Wilhelmshaven* (Berlin: Bundesanstalt für Materialprüfung).

Battelle (1978) *Risk Assessment Study for the Harbor of Gothenburg*; and *Risk Assessment Study for an Assumed LNG Terminal in the Lysekil Area*, prepared for the Swedish Energy Commission (Frankfurt: Battelle-Institut EV).

Behn, R.D. (1979) *Policy Analysis and Policy Politics* (Durham, NC: Institute of Policy Sciences and Public Affairs, Duke University).

Binnenlandse Zaken (1980) *Elk kent de laan die derwaart gaat*, Report No. 3 ('s-Gravenhage: Commissie Hoofdstructuur Rijksdienst, Ministerie van Binnenlandse Zaken).

Blokker, E.F. (1981) "The use of risk analysis in the Netherlands" *Angewandte Systemanalyse* 2(4):1968–71.

Blount, S. (1983) *Facility Siting and Neighborhood Compensation* Senior Thesis, Dept of Civil Engineering, Princeton University (unpublished).

Boulding, K. (1983) *National Defense Through Stable Peace.* Lectures delivered at IIASA, June/July 1981 (Laxenburg, Austria: International Institute for Applied Systems Analysis).

Braybrooke, D. (1974) *Traffic Congestion Goes Through the Issue-Machine* (London: Routledge and Kegan Paul).
Braybrooke, D. (1978) "Policy formation with issue processing and transformation of issues", in Hooker, Leach, and McClennen (eds). *Foundations and Applications of Decision Theory* (Dordrecht, Holland: Reidel).
Braybrooke, D. and Lindblom, C. (1963) *A Strategy of Decision* (New York: The Free Press).
Brötz, W. (1978) *Sicherheitstechnisches Gutachten zum Planfeststellungsverfahren eines Schiffsanlegers vor dem nördlichen Teil des Voslapper Grodens im Norden von Wilhelmshaven* (Stuttgart: Institut für Technische Chemie der Universität Stuttgart).
Brötz, W. (1979) *Sicherheitstechnisches Gutachten zum Antrag auf Vorbescheid zur Errichtung und zum Betrieb eines LNG-Terminals in Wilhelmshaven der Deutschen Flüssigerdgas Terminal GmbH, Wilhlemshaven* (Stuttgart: Institut für Technische Chemie der Universität Stuttgart).
Calabresi, B. and Bobbitt, P. (1978) *Tragic Choices* (New York: Norton and Co.).
Calabresi, B. and Melamed, A. (1972) "Property rules, liability rules, and inalienability: One view of the cathedral" *Harvard Law Review* 85:1089–128.
CBS (1978) *Statistisch Zakboek 1978* ('s-Gravenhage: Centraal Bureau voor de Statistiek).
CBS (1979) *De Nederlandse energiehuishouding* ('s-Gravenhage: Centraal Bureau voor de Statistiek).
Chevarley, U. (1975) "Power plant insurance and the Price–Anderson Act" *Risk Management* April:15–20.
Clark University Hazard Assessment Group and Decision Research (1982) *The Nature of Technological Hazard* Draft paper.
Cobb, R. and Elder, C.D. (1972) *Participation in American Politics: The Dynamics of Agenda Building* (Baltimore: Johns Hopkins University Press).
Conrad, J. (1982) "Society and problem-oriented research: On the socio-political functions of risk assessment", in Kunreuther, H. (ed) *Risk: A Seminar Series* Collaborative Proceedings CP-82-S2 (Laxenburg, Austria: International Institute for Applied Systems Analysis).
Cotgrove, S. (1981) "Risk, value conflict and political legitimacy", in R. Griffiths (ed.) *Dealing with Risk: The Planning Management and Acceptability of Technological Risk* (Manchester, UK: Manchester University Press)
Council for Science and Society (1977) *The Acceptability of Risks* (Chichester, UK: Barry Rose).
Cremer and Warner (1977) *The Hazard and Environmental Impact of the Proposed Shell NGL Plant and Esso Ethylene Plant at Mossmorran, and Export Facilities at Braefoot Bay* Vols I and II (London: Cremer and Warner).
Cyert, R. and March, J (1963) *A Behavioral Theory of the Firm* (Englewood Cliffs, NJ: Prentice-Hall).
Daniels, E.J. and Anderson, P.J. (1977) "International LNG projects continue to progress as new plans evolve" *Pipeline and Gas Journal*, June.
Davis, L.N. (1979) *Frozen Fire: Where Will it Happen Next*? (San Francisco: Friends of the Earth).
DFTG (1978) *LNG Terminal Wilhelmshaven. Kurze Projektbeschreibung des LNG-Terminals in Wilhelmshaven nach §9 des Bundes-Immissionsschutzgesetzes* (Wilhelmshaven).
DGWE (1979) *Genehmigungsvorbescheid (§4ff Bundes-Immissionsschutzgesetz): Errichtung eines Erdgas-Terminals in Wilhelmshaven der Deutschen Flüssigerdgas Terminal GmbH–DFTG* (Oldenburg: Bezirksregierung Weser-Ems).
Douglas, J. (1983) "How actual political systems cope with the paradoxes of social choice", in *Social Choice and Cultural Bias.* Collaborative Paper CP-83-4 (Laxenburg, Austria: International Institute for Applied Systems Analysis).

Douglas, M. (1972) "Environments at risk", in J. Benthall (ed) *Ecology: The Shaping Inquiry* (Harlow, UK: Longman).
Douglas, M. (1978a) *Purity and Danger* (London: Routledge and Kegan Paul).
Douglas, M. (1978b) "Cultural Bias" *Occasional Paper No. 35* (London: Royal Anthropological Institute).
Douglas, M. (1982) "Perceiving low probability events", in *Social Choice and Cultural Bias.* Collaborative Paper CP-83-4 (Laxenburg, Austria: International Institute for Applied Systems Analysis).
Douglas, M. and Wildavsky, A. (1982) *Risk and Culture: An Essay on the Selection of Technical and Environmental Dangers* (Berkeley: University of California Press).
Drake, E.M. and Kalelkar, A.S. (1981) "Handle with care: Using risk analysis for hazardous materials facilities" *Risk Management* (March) pp44—50.
Edwards, W. (1981) "Reflections on and criticisms of a highly political multi-attribute utility analysis", in L. Cobb and R. Thrall (eds) *Mathematical Frontiers of the Social and Policy Sciences* (Boulder, CO: Westview Press) pp157—9.
Edwards, W. and von Winterfeldt, D. (1981) *Towards Understanding Social Opposition to Risky Technologies* NSF Grant Division of Policy Research and Analysis.
Emory, C. and Niland, P. (1968) *Making Management Decisions* (Boston: Houghton Mifflin).
Energy Analysts (1978) *Eine kritische Betrachtung der geplanten Übergabe- und Transporteinrichtungen für den DFTG-LNG-Terminal in Wilhelmshaven, Deutschland* (prepared for Mobil Oil AG, FRG).
Fairley, W.B. (1981) "Assessment for catastrophic risks" *Risk Analysis* 3:197—204.
Farmer, F. (1976) "Safety of nuclear power in a modern society" *Riv. Int. Ecol.* 23 (6):595—604.
FERC (1978) *Final Environmental Impact Statement: Western LNG Project* Vol. III, Comments and Appendices FERC/EIS-0002F (Washington, DC: Federal Energy Regulatory Commission).
Fife, Dunfermline, and Kirkcaldy District Councils (1977) *An Assessment of the Shell and Esso Proposals for Mossmorran and Braefoot Bay.*
Fischhoff, B., Lichtenstein, S., Slovic, P., Keeney, R., and Derby, S. (1981b) *Acceptable Risk* (Cambridge, UK: Cambridge University Press).
Fischhoff, B., Slovic, P., and Lichtenstein, S. (1981a) "Lay foibles and expert fables in judgments about risk", in T. O'Riordan and R.K. Turner (eds) *Progress in Resource Management and Environmental Planning* Vol. 3 (Chichester, UK: Wiley).
Flint, A.R. (1981) "Risks and their control in civil engineering" *Proceedings of the Royal Society A* 376:1764.
FPC (1976) *Pacific-Indonesia Project: Final Environmental Impact Statement*, Bureau of Natural Gas, FERC (Washington, DC: Federal Power Commission).
Gamble, D. (1978) "The Berger Inquiry: An impact assessment process" *Science* 199: 946—52.
Gasunie (1978a) *Jaarverslag 1977* (Groningen: NV Nederlandse Gasunie).
Gasunie (1978b) *Memorandum aan de bijzondere commissie voor het stuk 14626: Aanvoer van vloeibaar aardgas (LNG) in Nederland; de keuze van de aanlandingsplaats; het Gasunie-standpunt* (Groningen: NV Nederlandse Gasunie).
Gershuny, J. (1981) "What should forecasters do? A pessimistic view" in P. Bähr and B. Wittrock (eds) *Policy Analysis and Policy Innovation: Patterns, Problems and Potentials* (London: Sage).
Groningen (1978a) *NOTA van gedeputeerde staten aan provinciale staten inzake de aanlanding van vloeibaar aardgas in de Eeemshaven* No. 56/1978; and *Nadere Nota van gedeputeerde staten aan provinciale staten inzake aanlanding van vloeibaar aardgas in de Eemshaven,* No. 56a/1978 (Provincie Groningen).
Groningen (1978b) *Aanlanding vloeibaar aardgas in de Eemshaven* Provinciale Staten vergadering van 25 mei 1978 (Provincie Groningen).

Guggenberger, B. (1980) *Bürgerinitiativen in der Parteiendemokratie* (Stuttgart: Kohlhammer).
Häfele, W. (1974) "Hypotheticality and the new challenges: The pathfinder role of nuclear energy" *Minerva* 1:303—23
Häfele, W. (1981) *Energy in a Finite World: Paths to a Sustainable Future* (Cambridge, MA: Ballinger)
Hirschman, A. (1970) *Exit, Voice and Loyalty: Responses to Decline in Firms, Organizations and States* (Cambridge, MA: Harvard University Press).
Holdren, J.P., Anderson, K., Gleick, P., Mintzer, I., Morris, G., and Smith, K. (1979) *Risk of Renewable Energy Resources: A Critique of the Inhaber Report.* Contract No: W-7405-EWG-45 (Washington, DC: US Dept of Energy).
Holling, C.S. (1981a) *Science for Public Policy: Highlights of Adaptive Environmental Assessment and Management* (Vancouver, BC: Institute of Resource Ecology) R-23.
Holling, C.S. (1981b) *Resilience in the Unforgiving Society*, Lindberg Lecture (Vancouver, BC: Institute of Resource Ecology) R-24.
HSE (1978a) *An Assessment of the Hazard from Radio Frequency Ignition at the Shell and Esso sites at Braefoot Bay and Mossmorran, Fife* (London: Health and Safety Executive).
HSE (1978b) *Canvey: An Investigation of Potential Hazards from Operations in the Canvey Island/Thurrock Area* (London: Health and Safety Executive).
Humphreys, P. (1982) "Value structure underlying risk assessments," in H. Kunreuther (ed) *Risk: A Seminar Series* Collaborative Proceedings CP-82-S2 (Laxenburg, Austria: International Institute for Applied Systems Analysis).
ICONA (1977) *Beleidsadvies bij het 'Rapport van de Projectgroep LNG Terminal'*, Report to the cabinet No. 84 ('s-Gravenhage: ICONA).
ICONA (1978a) *Nader advies van de ICONA inzake de aanvoer van vloeibaar aardgas (LNG) in Nederland*, Report to the cabinet No. 49 ('s-Gravenhage: ICONA).
ICONA (1978b) *Aanvullend advies van de ICONA inzake de mogelijkheid van aanlanding van vloeibaar aardgas (LNG) in het Eemshavengebied*, Report to the cabinet No. 145 ('s-Gravenhage: ICONA).
ICONA (1978c) *Jaarverslag ICONA 1978—1979* ('s-Gravenhage: ICONA).
Inglehardt, R. (1977) *The Silent Revolution* (Princeton, NJ: Princeton University Press).
Inhaber, P.H. (1979) "Risk with energy from conventional and nonconventional sources" *Science* 203:718—23.
Jennergren, L.P. and Keeney, R.L. (1979) *Risk Assessment* IIASA draft paper (unpublished).
Johannsohn, G. (undated) *Gefährdung der Umwelt durch gefährliche Stoffe bei Kollissionsfällen von Gas- und Chemikalientankern* (Bremen).
Johnston, R. (1980) "The characteristics of risk assessment research," in J. Conrad (ed) *Society, Technology and Risk Assessment* (London: Academic Press) pp105—22.
Jungermann, H., von Winterfeldt, D. and Coppock, R. (eds) (1982) *Analysis, Evaluation and Acceptability of Hazardous Technologies and their Risks: A Workshop Report* (Berlin: International Institute for Environment and Society, Science Center) IIES-dp82-2.
Keeney, R.L. (1980) *Siting Energy Facilities* (New York: Academic Press).
Keeney, R., Kulkarni, R., and Nair, K. (1979) "A risk analysis of an LNG terminal" *Omega* 7:191—205.
Keeney, R. and Raiffa, H. (1976) *Decisions with Multiple Objectives* (New York: Wiley).
King, R., and Nugent, N. (eds.) (1979) *Respectable Rebels: Middle Class Campaigns in Britain in the 1970s* (London: Hodder and Stoughton).

Kitschelt, H. (1980) *Kernenergiepolitik: Arena eines gesellschaftlichen Konfliktes* (Frankfurt/New York: Campus).

Krappinger, O. (1978a) *Abschätzung des Risikos, daß Tanker mit gefährlicher Ladung im Jadefahrwasser mit anderen Schiffen kollidieren oder auf Grund laufen* (Hamburg: Hamburgische Schiffbau-Versuchsanstalt GmbH).

Krappinger, O. (1978b) *Risikoanalyse: Über die Gefährdung der an den Umschlagsbrücken der DFTG und ICI leigenden Schiffe durch den die Anlage passierenden Verkehr auf dem Fahrwasser der Jade* (Hamburg: Hamburgische Schiffbau-Versuchsanstalt GmbH).

Krappinger, O. (1978c) *Ergänzung der Risikoanalyse: Über die Gefährdung der an den Umschlagsbrücken der DFTG und ICI liegenden Schiffe durch den die Anlage passierenden Verkehr auf dem Fahrwasser der Jade* (Hamburg: Hamburgische Schiffbau-Versuchsanstalt GmbH).

Kunreuther, H. (ed) (1982) *Risk: A Seminar Series* Collaborative Proceedings CP-82-S2 (Laxenburg, Austria: International Institute for Applied Systems Analysis).

Kunreuther, H. and Lathrop, J. (1981) Siting hazardous facilities: Lessons from LNG *Risk Analysis* 1(4): 289—302.

Kunreuther, H., Lathrop, J., and Linnerooth, J. (1982) "A descriptive model of choice for siting facilities" *Behavioral Science* 27:281—97.

Kunreuther, H. C. and Ley, E.V. (eds) (1982) *The Risk Analysis Controversy: An Institutional Perspective*, Proceedings of a Summer Study held at IIASA, Laxenburg, Austria, June 1981 (Berlin: Springer-Verlag).

Kunreuther, H., Linnerooth, J., and Starnes, R. (eds) (1982) *Liquefied Energy Gases Facility Siting: International Comparisons*, Proceedings of an IIASA Task Force Meeting, 23—26 September 1980. Collaborative Proceedings CP-82-S6 (Laxenburg, Austria: International Institute for Applied Systems Analysis).

Lathrop, J. (1980) *The Role of Risk Assessment in Facility Siting: An Example from California* Working Paper WP-80-150 (Laxenburg, Austria: International Institute for Applied Systems Analysis).

Lathrop, J. (1981) *Decision-Making on LNG Terminal Siting: California, USA*, Case Study Draft Report (Laxenburg, Austria: International Institute for Applied Systems Analysis).

Lathrop, J. and Linnerooth, J. (1982) "The role of risk assessment in a political decision process", in P. Humphreys and A. Vari (eds) *Analysing and Aiding Decision Processes* (Amsterdam: North-Holland).

Lave, L. (1981) *The Strategy of Social Regulation: Decision Framework for Policy* (Washington, DC: The Brookings Institution).

Lawless, E.J. (1977) *Technology and Social Shock* (New Brunswick, NJ: Rutgers University Press).

Lévi-Strauss, C. (1966) *The Savage Mind* (London: Wiedenfeld and Nicholson).

Lewis, H.W. (Chairman) (1978) *Risk Assessment Review Group Report to the U.S. Nuclear Regulatory Commission* NUREG/CR-0400. (Washington, DC: US Nuclear Regulatory Commission).

Lichtenstein, S., Slovic, P., Fischhoff, B., Layman, M., and Combs, B. (1978) "Judged frequency of lethal events" *Journal of Experimental Psychology: Human Learning and Memory* 4:551—578.

Lindblom, C. (1959) "The science of muddling through" *Public Administration Review* 19:79—88.

Linnerooth, J. (1975) *The Evaluation of Life Saving: A Survey* Research Report RR-75-21 (Laxenburg, Austria: International Institute for Applied Systems Analysis).

Linnerooth, J. (1979) "The value of human life: A review of the models" *Economic Inquiry* 17:52—74.

Linnerooth, J. (1980) *A Short History of the California LNG Terminal* Working Paper WP-80-155 (Laxenburg, Austria: International Institute for Applied Systems Analysis).

Linnerooth, J. (1982) "Murdering statistical lives . . . ?", in M.W. Jones-Lee (ed) *The Value of Life and Safety* (Amsterdam: North-Holland).

Linstone, H. *et al.* (1981) "The multiple perspective concept with applications to technological assessment and other decision areas" *Technological Forecasting and Social Change* 20:275—325.

Lloyd, G. (1978) *Gutachten: Zur Sicherheit von LNG-Tankern* (Hamburg).

Lowrance, W. (1976) *Of Acceptable Risk: Science and the Determination of Safety* (Los Altos:Kaufman).

Luce, R.D. and Raiffa, H. (1957) *Games and Decisions* (New York:Wiley).

McFadden, D. (1975) "The revealed preference of a government bureaucracy theory" *Bell Journal of Economics* 6(2): 401—16.

Macgill, S.M. (1982) *Decision Case Study: Mossmorran—Braefoot Bay, United Kingdom* Collaborative Paper CP-82-40 (Laxenburg, Austria: International Institute for Applied Systems Analysis).

Macgill, S.M. and Snowball, D.J. (1982) *What Use Risk Assessment*? Working Paper 323 (Leeds, UK: School of Geography, University of Leeds). To be published in *Applied Geography.*

Majone, G. (1978) "The uses of policy analysis", in *Russel Sage Foundation: The Future and the Past: Essays on Progress and the Annual Report 1976—77* (New York: Russel Sage Foundation).

Majone, G. (1979) "Process and outcome in regulatory decisions" *American Behavioral Scientist* 22:561—83.

Majone, G. (1980) "An anatomy of pitfalls", in G. Majone and E. Quade (eds) *Pitfalls of Analysis* IIASA Series on Applied Systems Analysis No. 8 (Chichester, UK: Wiley).

Majone, G. (1984) *The Uses of Policy Analysis* (New Haven, CT: Yale University Press).

March, J.G. (1978) "Bounded rationality: Ambiguity and engineering of choice" *Bell Journal of Economics* 9:587—608.

March, J,G, and Simon, H.A. (1958) *Organizations* (New York: Wiley).

Mazur, A. (1980) "Societal and scientific cases of the historical development of risk assessment", in J. Conrad (ed) *Society, Technology, and Risk Assessment* (London: Academic Press) pp 151—8.

Mechitov, A.I. (1982) *A Descriptive Study of Gas Pipeline Route Selection in West Georgia, USSR* Working Paper WP-85-56 (Laxenburg, Austria: International Institute for Applied Systems Analysis).

Meltsner, A.J. (1980) "Don't slight communication: Some problems of analytical practice", in G. Majone and E.S. Quade (eds) *Pitfalls of Analysis* IIASA Series on Applied Systems Analysis No. 8 (Chichester, UK: Wiley).

Moss, T. and Lubin, B. (1980) "Risk analysis: A legislative perspective", in C.R. Richmond, P. Walsh, and E. Copenhaven (eds) *Health Risk Analysis* (Philadelphia, PA: The Franklin Institute Press).

Mulkay, M. (1979) *Science and the Sociology of Knowledge* (London: George Allen and Unwin).

Murphy, D. *et al.* (1979) *Protest: Grüne und Steuerrebellen: Ursachen und Perspektiven* (Reinbeck bei Hamburg: Rowohlt).

National Research Council Governing Board Committee on the Assessment of Risk (1981) *The Handling of Risk Assessments in NRC Reports*, Report to the Governing Board (Washington, DC: US National Research Council).

Nelkin, D. (1971) "Scientists in an environmental controversy" *Science Studies* 1:245—61.

Nelkin, D. (1975). "The political impact of technical expertise" *Social Studies of Science* 5:35–54.
Nelkin, D. (1981). "Some social and political dimensions of nuclear power: Examples from Three Mile Island" *American Political Science Review* 75: 132–42.
Neustadt, R. (1970) *Alliance Politics* (New York).
New Scientist (1982) "Nuclear watchdog decides how safe is safe", 18 Feb., p421.
NMAB (1980) *Safety Apsects of Liquefied Natural Gas in the Marine Environment*, National Materials Advisory Board Publication No. NMAB 354 (Washington, DC: US Department of Transportation).
Nowotny, H. (1982) "Sociological proposals—critical comments", in H. Jungermann, D. von Winterfeldt, and R. Coppock (eds) *Analysis, Evaluation and Acceptability of Hazardous Technologies and their Risks: A Workshop Report* (Berlin: International Institute for Environment and Society, Science Center) IIES-dp82-2 pp35–46.
NRC (1975) *Reactor Safety Study: An Assessment of Accident Risks in US Commercial Nuclear Power Plants* Rasmussen Report WASH-1400 (Washington, DC: Nuclear Regulatory Commission).
OECD/Interfutures (1979) *Facing the Future* (Paris: OECD).
O'Hare, M. (1977) "Not on my block you don't: Facility siting and the strategic importance of compensation" *Public Policy* 25:409–58.
O'Hare, M. (1981) "Information management and public choice" *Public Policy Analysis and Management* 1:223–56.
O'Hare, M., Bacow, L.S., and Sanderson, D. (1983) *Facility Siting and Public Opposition* (New York: Van Nostrand Reinhold).
Office of Technology Assessment (OTA) (1977) *Transportation of Liquefied Natural Gas* (Washington, DC: Office of Technology Assessment).
O'Riordan, T. (1981) "Societal attitudes and energy risk assessment", in G.T. Goodman, L.A. Kristoferson, and J. Hollander (eds) *European Transitions from Oil—Societal Impacts and Constraints on Energy Policy* (London: Academic Press).
Orr, D.W. (1977) "US energy policy and the political economy of participation" *Journal of Politics* 41(4): 1027–56.
Otway, H. and von Winterfeldt, D. (1982) "Beyond acceptable risk: On the social acceptability of technologies" *Policy Sciences* 14(3) June.
Otway, H.J., Pahner, P.D., and Linnerooth, J. (1975) *Social Values in Risk Acceptance* Research Memorandum RM-75-54 (Laxenburg, Austria: International Institute for Applied Systems Analysis).
Owen, D. (1982) "Spectral evidence: the witchcraft cosmology of Salem village in 1692", in Douglas, M. (ed) *Essays in the Sociology of Perception* (London: Routledge and Kegan Paul).
Pfenningsdorf, W. (1979) "Environment, damages and compensation" *American Bar Foundation* 2:347–448.
Philipson, L.L. (1978) "Safety of LNG systems" *Energy Sources* 4: 135–55.
Plott, C.R. and Levine, M.E. (1978) "A model of agenda influence on committee decisions" *American Economic Review* 68:146–50.
Quade, E.S. (1975) *Analysis for Public Decisions* (New York: Elsevier).
Quine, W.V. (1953) *From a Logical Point of View: Logico-Philosophical Essays* (Cambridge, MA: Harvard University Press).
Raiffa, H. (1980) *Science and Policy: Their Separation and Integration in Risk Analysis*, paper delivered at the 6th Symposium on Statistics and the Environment, 6 October.
Raiffa, H. (1982) *The Art and Science of Negotiation* (Cambridge, MA: Harvard University Press).
Ramsay, W. (1981) "On assessing risk" *Resources* No. 68, October, pp10–17.
Rasbash, D.J. and Drysdale, D.D. (1977) *Fire and Explosion Hazard to Dalgety*

Bay and Aberdour Associated with the Proposed Fife NGL Plant (University of Edinburgh: Department of Fire Safety Engineering).

Reichel, P. (1981) *Politische Kultur der Bundesrepublik* (Opladen: Leske und Budrich).

Reijnders, L. (1982) "Societal interest", in H. Kunreuther, J. Linnerooth, and R. Starnes (eds), *Liquefied Energy Gases Facility Siting: International Comparisons*, Proceedings of an IIASA Task Force meeting, 23–26 September 1980. Collaborative Proceedings CP-82-S6 (Laxenburg, Austria: International Institute for Applied Systems Analysis).

Rijnmond (1977) *Notitie inzake diverse aspecten verbonden aan de inspraakprocedure m.b.t. de eventuele aanlanding van LNG op de Maasvlakte* (mimeo) Openbaar Lichaam Rijnmond (afd. milieubeheer) Rotterdam, 7 December.

Rijnmond (1978a) *Nota inzake de aanlanding van LNG* (policy statement, mimeo) Openbaar Lichaam Rijnmond 1978/613, Rotterdam, 7 June.

Rijnmond (1978b) *Statement by Rijnmond Public Authority to Parliamentary Committee 14626* (mimeo) Openbaar Lichaam Rijnmond, Rotterdam, 3 October.

Risikoabschätzung (1979) *Risikoabschätzung für den Transport und Umschlag von tiefkalt verflüssigtem Erdgas (LNG) und Chemikalien auf der Jade* (Bonn: Arbeitsgruppe des Beirates für die Beförderung gefährlicher Güter beim Bundesverkehrsministerium).

Ronge, V. (1980) "Theoretical concepts of political decision-making processes", in J. Conrad (ed) *Society, Technology and Risk Assessment* (New York: Academic Press) p209.

Ronge, V. (1982) "Risks and the waning of compromise in politics", in H.C. Kunreuther and E.V. Ley (eds) *The Risk Analysis Controversy: An Institutional Perspective*, Proceedings of a Summer Study held at IIASA, Laxenburg, Austria, June 1981 (Berlin: Springer-Verlag).

Rotterdam (1977) *LNG Aanvoer via Rotterdam* (Rotterdam: Havenbedrijf der Gemeente).

Rotterdam (1978a) *Aanlanding LNG, Verzameling gedrukte stukken 1978* Vol. 174 (75/47): 1019–62 (B&W Rotterdam).

Rotterdam (1978b) *Notulen Raadsvergadering* (Gemeenteraad van Rotterdam), pp215–243.

Rowe, W.D. (1977) *An Anatomy of Risk* (New York: Wiley).

Rowe, W.D. (1980) "Risk assessment approaches and methods", in J. Conrad (ed) *Society, Technology and Risk Assessment* (London: Academic Press).

Saaty, T.L. (1980) *The Analytic Hierarchy Process* (New York: McGraw-Hill).

SAI (1975a) *LNG Terminal Risk Assessment Study for Los Angeles, California*, prepared for Western LNG Terminal Compnay, SAI-75-614-LJ (La Jolla, California: Science Applications, Inc.) Chapters 1, 8, & 9.

SAI (1975b) *LNG Terminal Risk Assessment Study for Oxnard, California*, prepared for Western LNG Terminal Company, SAI-75-615-LJ (La Jolla, California: Science Applications, Inc.).

SAI (1976) *LNG Terminal Risk Assessment for Point Conception, California*, prepared for Western LNG Terminal Company (La Jolla, California: Science Applications, Inc.).

Scharpf, F.W., Reissert, B., and Schnabel, F. (1976) *Politikverflechtung: Theorie und Empirie des Kooperativen Föderalismus in der BDR* (Kronberg: Scriptor-Verlag).

Schulze, W. and Kneese, A. (1981) "Risk in benefit—cost analysis" *Risk Analysis* 1(1).

Schwarz, M. (1982) *Decision Making on LNG Terminal Siting: The Netherlands* Collaborative Paper CP-82-45 (Laxenburg, Austria: International Institute for Applied Systems Analysis).

SES (1976) *Environmental Impact Report for the Proposed Oxnard LNG Facilities* Draft EIR, Appendix B (Los Angeles, California: Socio-Economic Systems, Inc.).

SES (1977) *Survey of LNG Risk Assessment*, prepared for California Public Utilities Commission, Contract SA-31 (Los Angeles, California: Socio-Economic Systems, Inc.).

Simon, H. (1967) "The changing theory and changing practice of public administration", in I. de Sola Pool (ed) *Contemporary Political Science: Toward Empirical Theory* (New York: McGraw-Hill).

Simon, H. (1978) "Rationality as process and as a product of thought" *American Economic Review* 68:1–16.

Simon, H. (1979) "Rational decision making in organizations" *American Economic Review* 69: 493–513.

Slovic, P., Fischhoff, B., and Lichtenstein, S. (1979) "Rating the risks" *Environment* 21:14–20.

Slovic, P., Fischhoff, B., and Lichtenstein, S. (1983) "Characterizing Perceived Risk", in R.W. Kates and C. Hohemser (eds) *Technological Hazard Management* (Cambridge, MA: Oelgeschlager, Gunn and Hain).

Southern, D. (1979) "Germany", in F.F. Ridley (ed) *Government and Administration in Western Europe* (London: Martin Robertson).

Stallen, P.J.M. and Tomas, A. (1981) *Psychological Aspects of Risk: The Assessment of Threat and Control.* Paper presented at the International School of Technical Risk Assessment, Erice, Sicily, 20–31 May.

Starr, C. (1969) "Societal benefits vs. technological risk" *Science* 165:1232–8.

Steiger, H. and Kimminich, O. (1976) *The Law and Practice Relating to Pollution Control in the Federal Republic of Germany* (London: Graham and Trotman).

Stoto, M. (1982) *What to do when the Experts Disagree* Working Paper WP-82-65 (Laxenburg, Austria: International Institute for Applied Systems Analysis).

STUNET (1977) *LNG Terminal in de Noordzee* Projectgroep LNG Terminal ('s-Gravenhage: STUNET).

Thompson, M. (1979) *Rubbish Theory* (London: Oxford University Press).

Thompson, M. (1980a) *Political Culture: An Introduction* Working Paper WP-80-75 (Laxenburg, Austria: International Institute for Applied Systems Analysis).

Thompson, M. (1980b) *An Outline of the Cultural Theory of Risk* Working Paper WP-80-177 (Laxenburg, Austria: International Institute for Applied Systems Analysis).

Thompson, M. (1980c) "Fission and fusion in nuclear society" *Rain*, Newsletter of the Royal Anthropological Institute, London.

Thompson, M. (1981) *Beyond Self-Interest: A Cultural Analysis of a Risk Debate* Working Paper WP-81-17 (Laxenburg, Austria: International Institute for Applied Systems Analysis).

Thompson, M. (1982) *Among the Energy Tribes: The Anthropology of the Current Policy Debate* Working Paper WP-82-59 (Laxenburg, Austria: International Institute for Applied Systems Analysis).

TNO (1976) *Evaluatie van de gevaren verbonden aan aanvoer, overslag en opslag van vloeibaar aardgas* (Apeldoorn: TNO Bureau Explosieveiligheid).

TNO (1978) *Evaluatie van de gevaren verbonden aan aanvoer, overslag en opslag van vloeibaar aardgas met betrekking tot een Eemshaventerminal* (Rijswijk: TNO Bureau Explosieveiligheid).

TNO (1980) *De besluitvorming rond de aanlanding van LNG in Nederland* ir. H. van Amerongen (mimeo) (Apeldoorn: TNO Werkgroep Industriele Veiligheid).

Tuller, J. (1978) *The Scope of Hazard Management Expenditure in the US* (mimeo) (Worcester, MA: Clark University).

TÜV (1979) *Sicherheitstechnisches Gutachten im Rahmen des Planfeststellungsverfahrens nach dem Wasserstraßengesetz für die Errichtung und den Betrieb einer Transport- und Umschlagbrücke vor dem nördlichen Teil des Voslapper Grodens in Wilhelmshaven* (Hamburg: Technischer Überwachungsverein Norddeutschland eV).

Tversky, A. and Kahneman, D. (1974) "Judgment under uncertainty: Heuristics and biases" *Science* 185:1124—31.

Tweede Kamer (1974) *Energienota* (Government policy paper on energy) zitting 1974—1975, 13122, Nos. 1—2 ('s-Gravenhage: Tweede Kamer der Staten-Generaal).

Tweede Kamer (1978) *Rapport Onderzoek aanvoer vloeibaar aardgas (LNG) in Nederland* zitting 1977, 1978, 1979, 14626, Nos. 1—33 ('s-Gravenhage: Tweede Kamer der Staten-Generaal).

Tweede Kamer (1978/1979) *Handelingen* (Parliamentary debates) zitting (1978—1979) Nos. 5 & 6, ('s-Gravenhage: Tweede Kamer der Staten-Generaal).

US General Accounting Office (1978) *Need to Improve Regulatory Review Process for Liquefied Natural Gas Imports*, Report to the Congress ID-78-17 (Washington, DC).

Vahrenholt, F. (1980) "Chemieanlagen sind kein Schokoladefabriken!", in E.R. Koch and F. Vahrenholt (eds) *Im Ernstfall hilflos Katastrophenschutz bei Atom- und Chemieanlagen* (Köln: Kiepenhauer und Witsch).

Vaupel, J. (1981) *Analytic Perspective on Setting Environmental Standards*, draft report prepared for the Office of Air Quality Planning and Standards (Washington, DC: US Environmental Protection Agency).

Vaupel, J. (1982a) Statistical Insinuation and Implicational Honesty. *Journal of Policy Analysis* 1(2):261—3.

Vaupel, J. (1982b) "Truth and consequences: Roles for analysts and scientists in health, safety, and environmental policy making", in W.A. Magrat (ed) *Improving Environmental Regulation* (Cambridge, MA: Ballinger).

Vaupel, J., Kunreuther, H., Linnerooth, J., and Stoto, M. (1982) *An Abuser's Guide to Risk Assessment*, paper presented at Conference on Low-Probability/High-Consequence Risk Analysis, Arlington, VA, June (unpublished).

Walker, J. (1977) "Setting the agenda in the US Senate: A theory of problem selection" *British Journal of Political Science* 7:423—45.

Weinberg, A.M. (1972) "Science and trans-science" *Minerva* 10:209—22.

Weinberg, A.M. (1982) "Reflections on risk assessment" *Risk Analysis* 1(1) March.

Wildavsky, A. (1964) *The Politics of the Budgetary Process* (Boston: Little, Brown and Co).

von Winterfeldt, D. and Edwards, W. (1981) *Assessing Social Controversies about Energy Scenarios*, paper prepared for Workshop on Analysis, Evaluation and Acceptability of Hazardous Technologies and their Risks, Science Center, Berlin, 14—17 December.

Wittman, D. (1982) "Efficient rules in highway safety and sports activity" *American Economic Review* 72(1).

WSB (1978) *Transport gefährlicher Güter zu den geplanten DFTG/ICI-Umschlagsbrücken* (Aurich: Wasser- und Schiffahrtsdirektion Nordwest).

WSB (1979) *Planfeststellungsbeschluß für den Ausbau der Bundeswasserstraße Jade durch die Errichtung und den Betrieb einer Umschlaganlage für Flüssigerdgas der Firma Deutsche Flüssigerdgas Terminal GmbH—DFTG, Essen, vor dem nördlichen Voslapper Groden in Wilhelmshaven* (Aurich: Wasser- und Schiffahrtsdirektion Nordwest).

Wynne, B. (1978) "Nuclear debate at the crossroads" *New Scientist* 3:349—51.

Wynne, B. (1982) "Institutional mythologies and dual societies in the management of risk", in H.C. Kunreuther and E.V. Ley (eds) *The Risk Analysis Controversy: An Institutional Perspective*, Proceedings of a Summer Study held at IIASA, Laxenburg, Austria, June (Berlin: Springer-Verlag).

Zuid-Holland (1978a) *Aanlanding Vloeibaar Aardgas*, Gedeputeerde Staten van Zuid-Holland (provinciale staten vergadering) ('s-Gravenhage).

Zuid-Holland (1978b) *Notulen Provinciale Staten van Zuid-Holland—Vergadering 15 Juni 1978*. Provincie Zuid-Holland ('s-Gravenhage) pp 4349—71.

Subject Index

Name Index

About IIASA

The International Institute for Applied Systems Analysis (IIASA), a nongovernmental, multidisciplinary, international research institution, was founded in October 1972 by the academies of science and equivalent scientific organizations of 12 nations from both East and West. Its goal is to bring together scientists from around the world to work on problems of common interest, particularly those resulting from scientific and technological development. The present National Member Organizations of the Institute are:

The Academy of Sciences of the Union of Soviet Socialist Republics

The Canadian Committee for IIASA

The Committee for IIASA of the Czechoslovak Socialist Republic

The French Association for the Development of Systems Analysis

The Academy of Sciences of the German Democratic Republic

The Japan Committee for IIASA

The Max Planck Society for the Advancement of Sciences, Federal Republic of Germany

The Foundation IIASA — Netherlands

The National Committee for Applied Systems Analysis and Management, People's Republic of Bulgaria

The American Academy of Arts and Sciences

The National Research Council, Italy

The Polish Academy of Sciences

The Austrian Academy of Sciences

The Hungarian Committee for Applied Systems Analysis

The Swedish Council for Planning and Coordination of Research

The Finnish Committee for IIASA

About the Authors

Howard Kunreuther is Professor of Decision Sciences at the Wharton School, University of Pennsylvania, USA. He received his PhD in economics from MIT. His teaching and research interest are in the areas of decision processes, risk, and insurance, with emphasis on the application of theory to problems related to natural and technological hazards. He was leader of the IIASA Risk Group from 1980 to 1982, engaged in research on the decision processes associated with low-probability events, particularly with the implications of decision processes on public policy. He is the author of numerous publications in the fields of economics, management science, and policy analysis.

Joanne Linnerooth began working at IIASA in 1974 with an interdisciplinary group concerned with the public acceptance of nuclear power, and more generally with the economic and sociological issues of large-scale and potentially hazardous technologies. She studied industrial engineering at Carnegie-Mellon University and has a doctorate in economics from the University of Maryland. She has published numerous articles on the problem of evaluating public programs that may save or lose lives. Her interest in safety and risk issues has led her to more political interpretations of the use of quantitative techniques for public policy making. Presently she is working on a cross-national study of the institutional responses in various industrialized countries to the problems of handling hazardous chemical wastes.

John Lathrop is currently senior scientist at Woodward-Clyde Consultants in Walnut Creek, California. He is involved in a variety of projects, including a risk assessment of a hazardous waste incinerator, a large-scale facility siting study, and several projects aiding technology choice decisions at electric utilities. He is also developing a decision analytic approach to aid nuclear strategic force deterrence decisions, and is examining the role of decision analysis in expert systems for simulation.

Hermann Atz, Mag. rer. nat. (MSc), studied physics and mathematics at the University of Innsbruck, political sciences and statistics at the Institute of Advanced Studies, Vienna (associated with the research area of agrarian and regional policy), and at the University of Vienna. His fields of interest include technology and environmental policies, the impact of technological change on society, public acceptance and social control of science and technology, and peace research. From 1980 to 1982 he was affiliated with the Risk Group at IIASA, and is currently working as a consultant for the Office of Statistics and Research of the province of Bolzano, South Tyrol, Italy.

Sally Macgill is a lecturer in geography at the University of Leeds, UK. She is a graduate in mathematics and economics from this university and has a PhD in geography. She has teaching and research interests in the fields of energy resources, environmental problems and risk assessment, and longer-standing interests in spatial economic modeling and in mathematical analyses of social systems. The research that forms the basis for her contribution to this book was undertaken during spring 1981, during which time she was associated with the IIASA Risk Group.

Christoph Mandl received his PhD in mathematics at the Swiss Federal Institute of Technology, Zürich, and has an engineering degree in mathematics from the Technical University of Vienna. He was assistant professor for operations research at the Swiss Federal Institute of Technology, then chairman of the department of mathematics and computer science at the Institute for Advanced Studies, Vienna. He also spent a year as visiting scientist at the operations research center at MIT, and was visiting professor at the department for industrial engineering at Universidade Catolica in Rio de Janeiro. He is a lecturer in operations research at the University of Vienna and at present head of corporate planning at the Austrian company ELIN. Dr Mandl is the author of three books on operations research and forecasting. His field of interests includes strategic planning, network optimization, forecasting techniques, and risk analysis.

Michiel Schwarz is currently engaged in doctoral studies at the Department of Social and Economic Studies, Imperial College of Science and Technology, University of London. His work is concerned with public decision making on controversial technologies. In the last several years he has been involved in various research projects on science and technology policy at a number of institutions and organizations, including the Science Policy Research Unit of the University of Sussex, UK.

Michael Thompson is a cultural anthropologist, and he came to IIASA in 1980 to work on institutional aspects of risk management. Since 1982 he has been working on institutional settings and environmental policies. He attended Oxford University, England, received his PhD in anthropology from the University of London, and has been a lecturer and research associate at various European and American universities and research institutes. He is the author of *Rubbish Theory: The Creation and Destruction of Value*, and has published articles on aesthetics, the sociology of education, architecture, and the sociology of perception.